Is EVOLUTION True?

Why Darwin's Rejection of
Intelligent Design no longer makes sense.
Why Life really *is* a miracle
in the true sense.

A Ω

Dorrie

SOL OMNIBUS LUCET

Dedicated to all who keep their minds open,
in the scientific, philosophical and *spiritual* quests
to discover what may really be true. DG

"Miracles are *not* contrary to the laws of nature but only contrary
to what we *know* of the laws of nature." St. Augustine.

With grateful acknowledgement to all writers and publishers whose work is quoted or referenced here: John Lennox, Paul Davies, Hugh Ross, Sir Roger Penrose, James Shapiro, Matti Leisola, Esko Valtaoja, William Dembski, Jonathan Witt, Michael Denton, Michael Behe, Mark Eastman, Chuck Missler, Stephen Meyer, Antony Latham, Jonathan Wells, Chris Carter, Art Battson, Charles Darwin, Richard Dawkins, Michael Flannery, John Davidson, Luther Sunderland, Peter Tompkins, James Perloff, Dean Radin, Rupert Sheldrake, P. M. Atwater, Lyn Margulis, Michael Polanyi, Colin Patterson, V. J. Torley, James Tour, Neal Grossman, Sandie Gustus, Steve Taylor, Michael Egnor, Casey Luskin, Tom Bethell, John Morrison, Robert Shedinger, the Journal of the Discovery Institute, *Evolution News,* Dr Bradley Nelson, Anne Horne, Dr Mary Neal, Anita Moorjani, A.H. Almaas, and many others, too numerous to mention, who have conducted powerful, modern research into the *non-physical (spiritual)* aspects of reality over the last 150 years. Heartfelt thanks, too, to those have who put up with me on this subject over the decades, especially my dearest friend, T.

Copyright © 2025 Dorrie

All rights reserved. No part of this publication may be reproduced or transmitted in any form or by any means, electronic or mechanical including photocopying, recording or any information storage or retrieval system, without prior permission in writing from the publisher.

The right of Dorrie to be identified as the author of this work has been asserted by him in accordance with the Copyright, Designs and Patents Act 1988. First published in Great Britain in 2025 ISBN: 978-1-78963-535-5

You may contact author at isevolutiontrue@proton.me, although, with regret, replies cannot be guaranteed, due to time & resource limits. You may, also, find further materials regarding this work at isevolutiontrue.net, although, at time of publishing this is just an idea, so please forgive us if this takes more time.

Contents

Preface	4
Who is this book for?	5
Introduction	9
Evolution Questionnaire	23

IS EVOLUTION TRUE?

1.	Trouble with Creation	25
2.	Cosmic Accident	33
3.	Darwin's New Law	45
4.	Where's the Tree?	59
5.	Life's Explosion	71
6.	True Nature	83
7.	Fabulous Chance	87
8.	Copying Errors	96
9.	Cosmic Soup	107
10.	Darwin's Mistake	114
11.	Unintelligent Design	122
12.	Sudden Appearance	132
13.	Fabulous Upside Down Bats	140
14.	Science vs Materialism	148
15.	Iconic Fakes of Evolution	160
16.	Chance Builds Software?	171
17.	What is Life, Truly?	179
18.	Intelligent Design	183
19.	The Trouble with Hume	189
20.	Sorry Wikipedia, *No!*	195
21.	Evolution as Religion	201

BEYOND MATERIALISM   213

22.	No One to Know	214
23.	Spiritual Science	220
24.	Mind Beyond Matter	232
25.	Proof of Heaven	248
26.	The Great Mystery	255

Epilogue.	270
Further reading	276
Author's Answers to Evolution Questionnaire	278
End Notes	279

Preface

Do you believe in evolution or is that an odd question? Evolution is not something to *believe* in, but a simple fact, like electricity, or that our planet is round not flat? It's what I used to think. But, after my own time as a Darwin fan, I began to have my doubts, not about the roundness of our planet, but about his famous theory.

As I started to look closer, I was surprised to discover that his ideas are not nearly as well-founded as we have been led to believe. Was the Darwinian emperor *naked?* Had we been misled? Should he and his supporters have admitted, long ago by now, that his ideas did not explain life's history nearly as well as he, and they, first thought?

I also discovered that Darwin's ideas were very much influenced by materialism, the philosophy that only the physical universe is real. No *spiritual* causes, *Life and Mind* are purely physical things which can be explained in purely physical terms, including, even, our `immaterial thoughts and ideas, our loves and joys`.

However, for our theories to be scientific, they must be *testable,* otherwise they're just *philosophy.* So we will test Darwin's ideas on evolution, including the **materialist** thinking driving them, in light of much modern data of which he, and his steam age colleagues, knew very little.

We will also test the kind of *creationism* which tries to treat the bible, not merely as spiritual wisdom, but as if it is an early science text, which tells us that our earth is just a few thousand years old, and, all in it was created in just *six* literal days. We won't spend a lot of time on this second idea, as it is one which modern geology easily falsifies. Earth is clearly far older than a few thousand years. Instead, our two main questions will be:

(1) Is **Materialism** true? Are only the **physical worlds** real? No *spiritual* causes *within or beyond* them? No *spiritual* dimensions, no *Life* after death?

(2) Is **Darwin's theory** of Evolution true? The branching, **tree-like descent**, of *all* Life on Earth from just one or two ancestor microbe(s), via **natural selection**.

The answers to these questions have the deepest implications for our sciences, for our wider culture, for our very lives. We will try to answer them.

Who Is This Book For?

It is for you if you like to keep an *open* mind. It is for you if you ever wonder whether Darwin *really* explained Life's origins as is so often claimed? It is probably not for you if you'd rather just go along with current fashions or if you simply don't care.

But, if you *do* care, and you are open to new ideas, or to fresh looks at old ideas, this will, I hope, be of interest. For example, be you agnostic or *spiritually* inclined, you may have begun to notice, as I gradually did, various major things which Darwin's ideas did not explain. Important things which for reasons which we'll explore, the mainstream scientific community has either tended to ignore or to actively shut down.

If you are spiritually minded, you may like some good quality arguments to put to your friends who are fans of Darwin and his materialist (atheistic) philosophy and want to bring you over to their way of thinking.

On the other hand, I also write for you who have *no belief* that we live in a cosmos of intelligence and meaning. Whether you express your disbelief with the dry fire of an academic, the caustic mockery of a comic, or just with the worldweary sense that this Life, although it can be beautiful, can also be hard, and, perhaps, without purpose or meaning, I can understand.

We all have to face some illness, loss, and our eventual departures from this world. This can make it hard to believe in anything *'more,' behind* or *beyond* our familiar lives, let alone in a good and *positive* more.

Life's deeper sources, if real, cannot be *seen,* at least not physically, and, if there are *afterlife* realms, in subtle dimensions, *'next door,'* they are not readily discernible to most of us. I think this is one reason why so many of us have opted for Darwin's 'god-not-needed' theory of Life and evolution.

Not because his ideas bring us comfort, but because we believe them to be *true.* So to allow that he may have got natural history wrong, to concede that there may, after all, be *more* to life than mere chance—to create and to evolve all—could involve a considerable mental and emotional reset for us. To change our minds can be difficult and, for many today, Darwin did not merely provide us with an interesting theory on *Life's* origins, but with a complete worldview. Something along the lines of,

> "Like it or not, Darwin showed us that *Life* is just an accident which evolved by accident. Unlike our superstitious ancestors, we now realize that there is no *intrinsic* meaning to existence. While some of us still cling to the old beliefs in a miraculous 'something more,' which *just Is,'* to give rise to this world, and to *spiritual realms* beyond it, we are mistaken. Darwin, and science as a whole, have shown us that only the PHYSICAL UNIVERSE + MINDLESS CHANCE, which *just Is,* are truly real."

If this is your perspective, I can understand. Not least because, for a while, it was my own. If so, the chances of you reading this may be slim, as it is an argument against much of what you have believed to be true. But, if you do, and you are able to do so without throwing it straight in the trash, I bow to you. Why? Because it is not easy to consider ideas contrary to our beliefs, let alone to change our views.

I would, though, say that the science, which so many now say *supports* Darwin and his atheistic worldview, is often not as settled, or as uncontested, as we have been led to believe. As new evidence and new ideas become available, it is supposed to change its mind.

So, we'll consider some of the modern evidence, much of it unknown to Darwin and his 19th century friends, and do our best to follow it wherever it may lead, as all sincere thinkers are supposed to do.

Before we do so, though, it is fair to ask, of this, your very unknown author: "Why do you care so much about these subjects as to spend so many years, writing and re-writing this book, on aspects of which you are not even a specialist, let alone a recognized one?" I will try to answer.

As a child, I was not especially spiritual, but, we had religious assembly at school, and I believed in the simple way that children do. Then, as a teenager, I came across Darwin and his ideas. To my still fairly crude thinking, his theories seemed to provide explanations for life's deeper mysteries which made my prior, childlike beliefs in the *spiritual* unnecessary. This led me to become a convinced, Darwin admiring, teenage atheist.

Later, I learnt transcendental meditation, for its relaxation benefits, not for any spiritual reason. Yet, as I read more about TM and its background, famously taught to the Beatles by the Maharishi, I found myself more open, once again, to the classical idea that there might, after all, be *more* to existence than just the obvious **PHYSICAL DIMENSIONS** of reality.

Perhaps Hamlet was right. Maybe there were "more things in heaven and earth" than I had dreamt of in my Darwin admiring, young-atheist philosophy. It felt like a positive shift, with the sense, now, that there is more to Life than meets the purely physical eye. Existence was not, after all, as meaningless as my discovery of Darwin's theories had led me to believe.

Reading Yogananda's *Autobiography of a Yogi, Life After Life* by Raymond Moody, authors like Rudolf Steiner, Edgar Cayce, and Robert Monroe, discovering that one of the organizing principles of reality seems to be *not* just an arbitrary, one-off existence, but repeated physical lives, and a bespoke *spiritual* evolution for each of us, helped to make better sense of the data of our lives. We were all, it seemed, in a mysterious *soul school,* with fair, if firm rules. Two of which being *free will* and *karma.* We are free to think, to believe, and to act as we choose. With the proviso that, be it in this life, or in some future existence, we will reap as we sow—*karma.*

Of course, for the materialistically inclined, there *is* no *karma,* nor any deeper good. There are just a rAnd0m, and, fundamentally, meaningless Life and Death. This is one of the arguments of those, like Darwin in his day, and Dawkins in ours, who do not believe in the *spiritual.* They say,

> "If there really was a *deeper,* kinder, spiritual order to things, we'd be able to *'see'* it, in some way, wouldn't we? And, this strange world would not, surely, be so cruel and chaotic."

Their arguments used to be my arguments. But, among other things, they miss, I think, the value of *free will,* which the *spiritual worlds* seeming hiddenness afford us, and the chances for learning and growth this provides. After all, if everything went our way, every time, would Life not be pretty boring? As to nature's 'evils,' which caused Darwin some of his own spiritual doubts, we'll briefly address this topic towards the end of the book. While, surely they mean well, I think that various, famous, modern thinkers who argue, in the name of rational, logical science, that, at root, there is probably nothing more to reality, and to us, than MINDLESS **dust** + MINDLESS **natural laws** + MINDLESS **chance,** are doing humanity in general, and science in particular, a great disservice.

They maintain that 'the science' is on their side. That 'the science' has shown us that Life is just a *cosmic fluke* and that Darwin proved it. It is, more than anything else, this dubious claim, that 'science' is on the side of the MATERIALIST (atheistic) world view that caused me to write this book,

To challenge the conflation of science, which is supposed to be a dispassionate quest to **find out what is true,** with the *ideology* known as materialism. A belief system which, contrary to what many now believe, is neither scientific in principle, nor science as such, but, rather, an ideology about the basic nature of reality—which may or *may not* be true.

One of the core arguments, throughout this book, will be that materialism is *not* true, that it is possible to *demonstrate* that it is not true, that *spiritualism* makes much better sense of the data of reality overall, and that our understandings of the true mysteries of existence will become so much broader and more meaningful once we have consigned '**materialism**' to the dustbin of not merely bad, but false, ideas.

I can understand perfectly well, as perhaps you do too, why many, today, say they dislike the organized religions. Because these ancient traditions can be so dogmatic and oppressive. All too often, in history, it was their way or the highway, or worse. Sometimes this still happens, even today.

But to reject all notions of deeper meanings, all notions of the *spiritual* and the *super* natural, and to adopt **materialism** as our basic creed, just because we dislike the religions, and their officials, is far too high a price to pay for the resulting distortions in our understandings of reality.

This is why our realizing, once again, that there is more, *far* more to our existences, than thinkers like Darwin and Marx, in their day, and and Hawking and Dawkins in ours, have imagined, is, I think, incredibly important. Hence this book. I hope it may be of some service.

Introduction

Existence is a mystery. That there is *anything* at all, rather than *nothing* at all, is astonishing. She or he who says otherwise isn't paying attention. But, is the majestic, star spangled infinity, from which we briefly emerge, here, on our little planet, circling its star, mindless or meaningful?

Does Life arise due to *intelligent* causes, which, amazingly, mysteriously, *just Are,* or due to *mindless* causes, which, just as amazingly, and mysteriously, *just are?* And, can we work out, by various means, including scientific means, which of these two most basic ideas about reality is the more likely?

Until a century or so ago, almost everyone, rich and poor, well known or obscure, held to the *first* view. Life and Mind had *intelligent* causes. They were fascinating miracles of the *Unseen Causes* of things. Beliefs poetically expressed in the world's traditional creation stories. Visible nature derived, *ultimately,* from subtle, *spiritual* sources, from non-physical, *inner Nature.*

Reality was *multi-dimensional* and *inward* as much as OUTWARD. However, as modern science developed, and became much more effective at solving all kinds of problems, some began to dismiss the traditional creation stories, and to look for more scientific explanations for the facts of Life.

Facts like? Living things: what breathed Life into them? Did they have intelligent causes, or mindless? Magical birth. Cosmic miracle or cosmic accident? Mysterious death, meaningless end? Or sad, but meaning-imbued, transition to another *Life,* in *other* dimensions?

What, too, of mysterious soul qualities and conditions, like *Mind and Feelings, Love and Desire, Joy and Sadness, Curiosity and Creativity?* Did these fascinating *Soul* qualities have intelligent sources, or were they just accidents of a MATTER that was, itself, MINDLESS? What, too, of the plants and the animals, vertebrate and invertebrate, crawling and flying, water breathing and air breathing? The fascinating fossils? How did the species come to be? Did they arise all at once, or gradually? Universal common ancestor microbe to mighty spreading tree, as Darwin came to propose? Or were they more like a forest, many different trees, planted all at once, as some said, or at different times, as others maintained?

Were the species products of a mysterious, *Self-Existing, Intelligence, which* **Just Is**, *pervading all things,* from quark to galaxy? Or were they produced by wholly **mindless forces**? Deep questions, their answers so key to our understandings of *Who* and *What* we really are.

They were not, though, the kind of questions I gave a lot of thought to, when, as an inexperienced teenager, I first learned of Darwin, read about his famous voyage on *HMS Beagle,* and of his views which purported to answer some of these questions, and, as a result, lost all my prior beliefs in the more `subtle` causes of things.

As a not too deeply thinking teen, I swallowed Darwin's evolutionary scheme—where just one, or two, ancestral microbes had led, he said, via vast and incredible variations, to all today's species—without too much reflection. I concluded that he had found a wholly naturalistic explanation for Life which made belief in the *super* natural unnecessary.

I became a convinced teenage-atheist and considered all who still held religious or spiritual beliefs uneducated and 'unscientific.'

Yet, just a few years later, I began to have my doubts. There was so much Darwin's ideas did not explain, and I began to understand that there might be more, *far more,* to reality than he realized or wished to consider.

I also began to realize that what we *believe* about reality is, often a matter of what we *want* to believe, and what *parts* of the available evidence we are *willing* to consider, rather than solely a matter of science or religion. For some of us, no amount of evidence, if it conflicts with our *prior religious—scientific—or other—beliefs*, will cause us to change our minds.

No one can *'prove'* anything to us if we are not inclined to agree—not even that the moon is *not* made of cheese, or that the earth is *round,* not flat.

This may mean I have wasted my time, writing and re-writing this piece, for so many years now. Why? Because my original premise was that, "If I just present the evidence showing the (many) gaping holes in Darwin's theory in particular, and in materialism [atheism] in general, people will say, *"Well done, Dorrie!* Why didn't we think of that!'" What *self-delusion!*

But why the debate? Because *Life's* deeper sources are not *obvious*. If they're intelligent, they're *subtle*. They do not, surely, look like a cosmically scaled, Father Christmas god, with a beard, sitting on a far off cosmic cloud, creating and maintaining it all. No. They can only be *inferred*, using reason and LOGIC, or known, to some extent, by *intuitive or spiritual* means.

Today, however, it is a common view that the evidence for Darwin's theory is so obvious, that any notions of an *intelligently* driven unfoldment of Life are outdated. Yet, and this is the point, when looked at just a little closely, Darwin's ideas do *not* account for the data. It's a controversial claim I know, especially here, in the intellectual west, where those of us who question Darwin tend to be dismissed as cranks or religious literalists.

For a while, I too was one of the mockers. I regarded any who didn't believe in Darwin's gospel irrational. At the time of my teenage conversion to **materialism**, via Darwinism, which seemed so much more *realistic* and scientific to me than the simple *spiritual* gleanings of my childhood, I did not realize that just because the old creation stories were not literal descriptions of natural history, it did not mean they contained no truth at all.

For example, just because the simplistic creation story which goes,

'In the beginning were the wonderful Mr. Rolls and Mr. Royce, and they said, "Let there be Rolls Royces!" And so it *was!*'

is not an engineer's description of the famous cars genesis, it is perfectly true, in its own *symbolic* way. Perhaps, likewise for *Genesis* in the *Bible*.

'In the beginning God created the heaven and the earth. ... And God said, Let there be light: and there was light. ...'

These words are not *science,* but they may be *spiritually true*. On the other hand, physical science descriptions of the universe and its history are not intended to be *spiritual,* but to describe physical reality.

Describing the Big Bang and later events, clever as it is, simply describes a *sequence* of physical events. But such descriptions of MATTER and its MECHANISMS across time, do not, necessarily, amount to *full explanations*.

'At 0 seconds the Big Bang happened and the fundamental forces were unified. At (0 + a) seconds gravity separated, leaving the other forces unified. At (0 + b) seconds the stars appeared, at (0 + c) seconds Life on Earth, then, eventually, people appeared, at (0 + d) seconds People began to farm and to live in villages . . .'

Similarly, someone could describe an automated car making plant:

'At (0 + a) seconds the Chassis was made. At (0 + b) seconds the Engine was mounted. At (0 + c) seconds the Wheels were added...'

Neither of these depictions of a series of events explains *why* there should be a Big Bang, or a factory, followed, in the case of the universe, by the arising of amazingly precise physical laws, smart-materials elements, like CARBON and OXYGEN, or stunningly complex, DNA CODED, Life.

They don't tell us of the *inner* causes of things. Albeit, according to Darwin's view, my teenage view, today a very common view, there *are* no *inner* causes. No, all things, even *living* things, are products of mere chance. There's nothing *inner, or intelligent, or spiritual* to play any causal role.

But, is this right? Has science honestly shown it to be? Is physical nature the only kind? I think science can help us answer some of these questions. And the answers it gives us on origins are among the most important of all.

For, if we are in a wholly MATERIAL universe, with no *spiritual* dimensions, one consisting solely in rocks and stones, we are truly alone.

If so, our *caring hearts* and our *bright minds* must have emerged from the apparently LIFELESS ROCKS and MINDLESS LIQUIDS of our planet by pure chance. No other option. If there is, please let me know. This is why I write, because I think this idea is irrational, deeply. Yet it is mainstream today, especially among those who like to think they are being 'scientific'

Yet, historically, most of humanity's greatest minds always held that there is far *more* to us than mere chance. Some also held that it is not just *one* spiritual force, which gives rise to reality, but hierarchies of various spiritual forces—all, ultimately, emanating from a *Final Source and Uncaused Cause* of unimaginable beauty, power, intelligence and creativity.

'Oh, you mean *God?'* Well, some call *It* that. Others give *It/Is/S/He (ISH),* different names: *the Tao or Brahman, Allah or Buddha Nature, the Great Spirit or the Great Mystery.* But, while for some these words are interchangeable, for others, God is not the same as the Tao, and Brahman is not the same as Allah, because of their differing beliefs around these terms.

For our purposes though, the diverse views among the world's believers, as to the exact characteristics of the *Higher Power* they believe in, are not important because we'll be looking at a far more basic division. That between MATERIALISM and *spiritualism,* sometimes also called *idealism.*

'Materialism—sometimes called naturalism—says: "Only the physical is real." MATERIALISM *vs.* `Spiritualism,` naturalism *vs. super-naturalism.* MINDLESS Universe vs. **Intelligent**? Which is more likely *true?*

Is it true that Darwin, and other scientists, have discovered that, due to a very long series of very unlikely chemical accidents, non-living elements, like carbon, eventually became *alive* and cell shaped, then fish shaped, frog shaped, and, finally, people able to *think* and to *create,* for no reason at all?

Is *this amazing, subtle, mysterious, Intelligent, Curious, Creative, Awareness–Beingness,* which quietly looks out through our eyes, this *subtle, ever-witnessing Awareness* which we all share, mere coincidence? Is every spiritual experience, which whisper to us of the *Soul* and of the *multi-dimensional,* just a hoax or a hallucination? According to Richard Dawkins,

> "In a universe of electrons and selfish genes, blind physical forces and genetic replication, some people are going to get hurt, other people are going to get lucky, and you won't find any rhyme or reason in it, nor any justice. The universe that we observe has precisely the properties we should expect if there is, at bottom, no design, no purpose, no evil, no good, nothing but pitiless indifference." [1]

In such a universe, Darwin's theory, or something similar, is the only possibility. If there is no *God, Spirit, or Higher Nature,* of any kind, the only possible creative agency *has* to be pure, mindless, chance (PMC).

On the other hand, is there *a mysterious, Living, Intelligence* to all things, from atom to galaxy? Including one with which it is possible, even, to have an *inner communion* or *spiritual relationship* of some kind? For our materialist friends, the answer has to be "no." There is no 'intelligence' or 'meaning' to any of it. From atom to galaxy, it is all mindless and dead.

This work is our inquiry, into which of these two most basic hypotheses of reality is more likely right? Does the Darwin–Dawkins, MINDLESS *universe* hypothesis best describe reality?, or, *Spiritualism's Intelligent Universe* hypothesis? Which makes better sense of the data, overall?

Creationism and intelligent design (ID) are mocked by many today. But most critics of these classical ideas seem to find it difficult to think of existence in a more *subtle* way than is allowed either by religious creation stories, *taken too literally*, *or* by the totally materialistic, and random-chance-based, accounts of Life, which thinkers like Darwin provide. They make no allowances for the *subtleties* of spirit and the mysteries of creation.

For example, while the *Mysterious Source of All, call it Spirit,* can be experienced in *precious, personal, inner* ways, it also has terrifyingly powerful, impersonal aspects: implacable cosmic laws, violent creation and violent destruction. Stupendously powerful cosmic forces, in mysterious contrast with *Spirit's* tender, *inner* sides, deep paradoxes, to be sure.

But, for many modern thinkers, if the world's scriptures turned out not to be literally true, in every possible way, then, maybe, they could not be trusted at all, including when it came to creation. To parody:

> "Unlike our forbears, we are a scientific race. We cannot see *'God' or 'Spirit' or 'Souls'* in our telescopes. So, your imaginary *Higher Power or God-Force* or whatever else you want to call it, which **Just Is** cannot exist, can it?
>
> "Does **not** a MINDLESS, RANDOM, UNIVERSE, *which* **Just Is**, make better sense, overall, of the data of our own *clever* existences, our own *Loves,* our own *Willing,* our own *Creating,* our own *bright, Intelligences?"* Ah me...

For those opposed to my one long argument, herein expressed, it is very easy to say to us, as many have done: "Life, Everyone, is just an amazing series of coincidences, and Darwin proved it. Inanimate CHEMICALS naturally evolved into **Glorious Life,** and, eventually, into you and me."

'That's interesting,' we reply. 'Can you tell us *how* they did so?' "Well," they say, "it was just a matter of time and *probability*.

"There were CHEMICAL-SOUP seas, flashes of lightning, and, hey presto, *Life!* Beautiful, isn't it? It was all a *natural* process.

"Simple, single cells [which are not simple at all!] arose by lucky chemical chance. These asexual cells *evolved,* by more chance, into ever more complex, and, eventually, *sexually reproducing—multi-celled—*snails, worms, spiders, and fish.

"The water-breathing fish *naturally evolved* into amphibians with completely different breathing systems, then dinosaurs.

"The dinosaurs evolved into birds, with very different lungs again. Land mammals, like Bears, *evolved* into deep sea diving, echo-locating Whales. Crawling Caterpillars evolved into lovely, multi-hued Butterflies. Rodents, like shrews, *evolved* into amazing, fast-flying, sonar equipped Bats.

"Simple-minded apes *evolved* into calculus calculating, concerto composing, computer constructing people, for no reason or *'purpose'* at all. *Intelligence* played no part, at any stage, which is intriguing, especially as we are so *bright!* It was a totally MINDLESS EVOLUTIONARY PROCESS. Isn't it amazing?"

I hope you will forgive the gentle parody, but it is ironic that so many people today, both scientist and lay, have replaced an, at times, too literalistic religious creationism with such intellectually vacuous claims as these.

I feel embarrassed to think of the times when, as a Darwin admiring teenager, I, too, trotted out this kind of lazy, pseudo-scientific, nonsense.

With far too little reflection, they claim, as I used to claim, in the name of sober, logical science, that this or that stunningly complex Living thing, "naturally evolved, that no *intelligence* was involved," without possessing the knowhow to see if their claims have any *scientific* basis.

That these kinds of assertions can *easily* be falsified, that they *have* been falsified—not once but *repeatedly*—so many of us now ignore. Why, though?

Mostly, I think, because *Life's* deeper sources are not obvious, the religions can be far too judgmental, dogmatic and annoying, because *Life* can be hard, and because we grew up hearing that Darwin had solved Life's mysteries, and that the *spiritual* is make-believe. So any data showing that all this may *not* be true now makes little difference to our views. "Life," we say, "is just a long-running accident and Darwin proved it." Our confidence is not justified but it is our *view*. Others maintain that their religious beliefs, no matter how extreme, be accepted without question or reflection.

Many years after I left my teenage atheism behind me, one of the odd things I discovered, when looking more closely at this controversial topic, is that so *obvious* are the problems with Darwin's ideas, it is really very easy, even for non-scientists, to pick up on them. If this were not so I would not have attempted to tackle such a complex and oft contested subject.

As it happens, I have no problem with evolution as a useful word to describe various types of change—*r*A*n*d*o*m or **intelligent**—over time. But, did evolution, as *Darwin* described it, take place? Then, *if* it did, did it do so by the wholly rAnd0m means he argued for, or, was it an *intelligent* process?

There were, in fact, plenty of well qualified people in Darwin's day who were skeptical of his claims for a totally mindless, unguided, intelligence-not-needed form of evolution. They did not argue on religious grounds, but they said that the data, from the fossils, now much added to by modern data from genetics and molecular biology, did not support his scheme.

That his views seemingly prevailed, and his *scientific* critics were sidelined over time, says, I think, as much about our changing attitudes towards religion and the *super* natural over the last couple of centuries, as it does about his *actual* ideas, the *real* science, and the *real* data. Here you will hear from some of them, and their successors. They deserve to be heard.

Still, if it is so clear, by now, that Darwin's ideas are *not* supported by the data, and, strangely, they never were, why do more of us not question?

I think it's mostly because the 'experts' tell us that evolution is just a fact, like gravity, and because we are given the **false choice** between taking religious creation stories too literally or following Darwin's newer *"Just So"* story. We seem, though, to so hate not to *'know.'* So, rather than admitting that we still have no real idea, *scientifically* speaking, how *Life* begins and unfolds, how *breathing and feeding, feeling and needing* began, we cling to

Darwin's outdated ideas. It is also no help that for many academics today it can be professionally risky to question Darwin's ideas in particular, or materialism in general, let alone to show any interest in the *super* natural.

Darwin, many now believe, showed MATERIALISM to be the better view, and all notions of the spiritual to be *delusions*. It was, after all, the adoption of his ideas by highly influential intellectuals like Marx, which so greatly contributed to the collapse of religion and spirituality throughout the West and the wider world. *God, Spirit, Meaning,* were, for many, now dead.

This was not trivial. Every year, countless young people, with some belief in the *spiritual,* go to schools and colleges, all over the world. Then, just as I, and perhaps you, once did, they learn about Darwin's famous theory of evolution, and, in the name of *science,* this supposedly most logical of disciplines, they are taught to doubt whether there is anything more to reality than *blind* chance, *blind* laws, and MINDLESS MATTER.

Not only can this make life feel meaningless, what if it's *wrong?* Eventually, we all pass through the fascinating mystery we call death. Does it not matter, then, whether our living and our dying are part of something meaningful and ongoing, or, fundamentally, meaningless? Yet, in the science community today MATERIALISM is the mainstream view. Those scientists brave enough to question it tend to be ignored. But, for an idea, like Darwin's, to be *scientific,* IT MUST BE TESTABLE, otherwise it's just *ideology*.

Well, it *can* be tested, and it *has* been tested, very thoroughly by now. What is *not* yet much admitted, is that Darwin's famous theory has *failed* those tests, not once but repeatedly. It really surprised me, dear reader, as I researched these topics, to discover that even some of the world's **best qualified** fossil experts have **publicly admitted** that the **fossil, and much other data besides,** so key to Darwin's theorizing, has *never* supported his ideas!

A friend once mentioned physicist Richard Feynman's comment that he'd rather have questions in science that can't be answered, than *answers* that can't be questioned.[2] Unfortunately, Darwin's ideas, and the materialist philosophy driving them, have, for many today, become answers to Life's mysteries which can't be questioned, even when the evidence for Darwin's views in particular, and for materialism in general, doesn't stack up.

Our debates around origins can become very charged, part of the culture wars even, with some seeing *science* and *religion* as in a fight to the death. They portray 'science' as unfailingly on the side of logic, rationality, and evidence-based thinking, and religion as pure superstition and blind belief.

So, they say, either we must take the old religious stories as *literally* true, or we must choose *'reason'* and adopt a wholly **materialistic** scheme, like Darwin's. Either the *religious,* with their *'superstitions,'* are right, or the *materialists,* who claim to be the owners of *'reason'* and the *'scientific'* view, are right. It is a false choice with no middle ground. Whereas, actually, there is *a lot of middle ground.* We will steer towards it. We will, especially, and emphatically, be arguing that science and materialism are *not* the same thing. Science is supposed to be an *open minded inquiry* into reality. Materialism is just one particular belief system about the nature of reality.

One of the challenges, though, for those of us who question Darwin's anti-spiritualist theory of Life's history, is that we are trying to open up thinking in an area not vital to practical, everyday life—unless *spirituality* is already very important to us—yet one which involves some of the deepest questions we can think about.

Who are we? ***What*** are we? Is there any ***meaning*** to our existences? While, I no longer agree with Darwin or my old, teenage self, that we can explain *all* of *Life* without reference to *subtle,* non-physical forces, *within, above, or beyond,* obvious, outer nature, none of the arguments which follow are intended to give weight to any particular religion or spiritual tradition.

None, at least, beyond the classical contention that the *super* natural and the *spiritual* are real, alongside the claim that there are, today, plenty of good quality, non-religious, *data* which give credence to their existence.

Science, religion, and spirituality all make truth claims. But, while the religions can demonstrate some of theirs quite well, that it is better, for example, to be kind than to be cruel, they cannot prove many of their more *specific* tenets, because they relate to past events which cannot now be verified. This is where personal choice and religious faith come in.

Science, though, is not supposed to be a matter of *faith*—but to test its claims. The once crazy-seeming idea, made so fashionable by Darwin, that intelligent *Life,* and intelligent *Mind,* may just be chemical accidents, which evolved by accident, needs to be tested, and tested hard, for it had—it *still* has—such massive, emotional, cultural, and political implications.

If true, it makes our lives, in essence, *meaningless.* If false, Life becomes so much broader and more fascinating to explore once more.

Luckily, the flame to know what is true burns bright in us. It burned *bright* in Darwin. It burns *bright* in Dawkins. It burns *bright* in any true thinker. But *bright* and *right* are not always the same. We can *believe* what we like, be we *'bright'* or not. That the earth is flat, that the *bright* moon is made of *bright* cheese, or that *bright Mind* could arise by mindless chance.

I'm curious, though, as to how many contemporary materialists and atheists including some very famous ones, ever stop to wonder **why** they **love truth,** or *what* it is, *in* them, which so tirelessly pursues the *truth,* as they see it? What, after all, do the MINDLESS ATOMS of the materialist faith care about *truth?* Why would they ever become *Alive,* or capable of thought, let alone interested in such *immaterial* things as *Beauty, or Goodness, or Truth?*

"Oh, just dumb chance, which **just is,**" many now say. But is this, *honestly,* an intellectually satisfying answer? Especially, when it is a demonstrable non-starter, not only in logic, but in strictly *scientific* terms.

That there are *any things-at-all,* like the laws of nature, like plants, or people, planets or stars—rather than no-things-at-all, is the *deepest mystery* of all. *Why? Because,* **no one***,* no matter how clever, can say **why** there is any-thing-at-all, rather than no-thing-at-all. Nor can anyone tell us *why* there is an ultimate *Cause of Things,* be it MINDLESS, or be it *intelligent,* not, at least, without falling into an infinite regress. For example, someone says:

"Everything exists because of *Cause 'Z,'* be it that cause *'Z'* is mindless, or, be it that it is *highly intelligent*—take your pick, which ever you prefer."

"What makes you say that?"

"Because *Cause 'Z'* caused the amazing, blazing, Big Bang, which led to everything else." "Ok. But what caused *Cause 'Z'* to so mysteriously *be?"* "Well, we think cause *'Y'* caused cause *'Z'* which caused the Big Bang." "Hmm, that's interesting. So what caused *'Y'?"*

"Maybe *'X'* caused cause *'Y,'* which caused *'Z,'* which caused the stupendously, astonishingly, staggeringly big, Big Bang."

"So what caused *cause 'X'. . ?"* The same question repeating for ever.

This is why we will never be able to say **why** there is an ultimate *Source* of things, whatever, exactly, *It/S/he or ISH* is, which *'just Is.'* No, only that, **we *know that Existence and its Source exist***—because *We* exist, because our *own, Living, Knowing, Beingness,* **knows *Living, Knowing, Beingness.***

Unlike our plant and animal brethren, who, as far as we can tell, do not have this fascinating possibility, we, *strange,* mysterious, *feeling,* thinking, *creating,* human beings, *can* try to work out the *true nature* of this *amazing,* **Aware, Living, Intelligent, Beingness**, as it arises *in us, and around us, 24/7,* and of its unfathomably mysterious *Source,* which *'**Just Is**,'* and whether it is, in any ways, like its creations, and its expressions, including we.

Is it inherently *intelligent,* as most of humanity has held? Or, is it inherently MINDLESS, as thinkers like Darwin and Dawkins have argued, but, due to *mindless* chance, it *mindlessly* gives rise to clever, *mindful* us.

I hope that, so far, all this sounds reasonable enough. However, it is fair to ask, of this, your unknown author, why should you trust my ideas more than those of world famous thinkers like Darwin in his day, and Dawkins in ours?

Firstly, because I have carefully researched this topic, and, secondly, because I rely, much, on the works of those—*scientists especially*—who are *authorities,* often world class ones. So if my *scientific* arguments have any merits, it is due to their expertise, not to mine.

Now, in theory, when we discover *new information,* or new ways of looking at old information, it should be easy for us to change our ideas. Why after all, would we not *gladly* leave behind the outworn and the outdated? Better information will, surely, lead to better ideas. Ah, if only it were so!

Why? Because, often, we have huge resistance to changing our views. For example: "Most scientists say Darwin's theory is true, so it must be true!" Or "Democracy, or Capitalism, or Socialism, or Materialism, or our Religion, or some other "ism" is the right way and yours is the wrong way!"

Oddly, we are often more flexible when it comes to new, *practical* ideas, like adopting a new technology, than modulating our *beliefs.* For example, right now, we have the widely shared *cultural* assumption that Darwin's theory is based on some very good—and well proven—science. However, this now widely held *faith* in Darwin's ideas—and in the atheistic philosophy driving them—is based on some very *questionable* assumptions which, all too often, go unnoticed. So, dear reader, this is where I hope you will consider looking at a few questions, just two pages on from here, on page 21, designed to stimulate our thinking on some of these, all too often, unquestioned assumptions, before we get more fully into our inquiry.

This is an unusual idea, I know, but these topics do require some thought. Whereas, quite often, I think, we would rather just stick to what we already know, or *think* we know, rather than do any real thinking of our own.

There is an added, but very important nuance to this *entire* topic. It is that our present science of *Life's origins*—because it doesn't affect whether a plane will safely fly or a drug will be safe—can adopt all kinds of ideas as 'truth' that would **not last five minutes** in other scientific fields.

It is these shoddy ideas, dressed up as science, many of them originating with Darwin, that I want your BS detectors to be open to, all the way through, as we repeatedly ask, which of the two most basic ideas on creation or origins—**intelligence** *vs* blind CHANCE—makes the best sense of the data overall? Does Darwin's **mindless universe hypothesis** make the best sense of the data? Or does Buddha, Plato, Jesus, Newton, Galileo's *intelligent, spiritual* universe hypothesis make the best sense of the data, overall?

You can, of course, skip these questions, or, perhaps, return to them later, when they may feel more meaningful, in light of the main text.

If, though, you do care to consider them, be it right away, or later, I will be honored. I may never know your answers, but my own, brief answers are at the back. My longer answer is, of course, this book as a whole.

These questions, and the issues they raise, we will return to again and again. Reduced to their most basic and their most essential, they are:
(i) Is **materialism** true? (ii) Is Darwin's **materialist** theory of **evolution** true?

To make these sometimes quite dry subjects more engaging, we will, metaphorically, join two friends as they walk and talk on a day out in the countryside. Alisha and Oliver will approach these ancient questions as to the *true nature* of reality, and *Life's true Origins,* in simple but radical ways.

They will question, as thoughtful children might, various *'truths'* of our current sciences and our wider culture. It was, after all, a child who pointed out that the foolish emperor in the *Emperor's New Clothes* story wasn't

21

wearing anything. "Is this," they will ask, "now also true of Darwin's theory in particular, and of materialism (atheism) in general? They may look quite good from a distance, but look, just a little closer, and they soon start to fall apart."

I hope you will forgive the simple style, including much EMPHASIS. We are encouraged not to '**shout**' in print, I know. But, I hope this may help to *AWAKEN!* us, who, like sad characters in a gloomy fairy tale, have fallen into such a dismal, Darwin induced slumber, regarding Life's *true* origins.

While this work can be read in one go, it can also be treated as a series of meditations: Could Life arise by *chance,* as Darwin believed? Did it do so? Is *materialism* scientific? Is *evolution,* as he described it, *true?*

I hope you will forgive the endless repetition of these key themes, but we will be questioning two now *very entrenched* and *mutually reinforcing* sets of beliefs about the basic nature of reality:

(1) **Materialism** or 'radical atheism' which has, by definition, to argue for a *non-Spiritual,* chance-based theory of life, be it Darwin's or someone else's.

(2) **Darwin's theory**, which, for many today, is one of their *key* reasons, as it once was for me, for adopting materialism as their basic philosophy in the first place. Darwin, many now think, showed us that *we do not need* to resort to religion or to the *super* natural to explain anything. Blind chance can create all.

Note: It is **materialist-atheism** or **radical atheism**—the belief that only the PHYSICAL is real—which this book questions. **Idealist-atheism,** by contrast, which allows for the *super* natural to be real, is not questioned here. Buddhism and Taoism are idealist-atheist *spiritual* traditions. They do not (officially) believe in a god or gods, but they do believe in the *super* natural, that is, in the **real existence** of *non-physical, spiritual, and spiritually causal* dimensions.

To confront and to disentangle these two now very entrenched sets of beliefs, **Darwinism** and **materialism**, is not easy. But by repeatedly revisiting these *key* ideas, we will, I hope, gradually highlight to any dear readers not initially persuaded, that we will have to rely, not on *science* as such, but on our *prior* ideas and our prior philosophies to disagree with the ideas herein mentioned. Why? Because, the **real science***,* despite all the claims for Darwin, despite all the assertions that: '**SCIENCE = MATERIALISM**' does not support their position. Well, it's time to look more closely.

I hope you enjoy the read.

Evolution Questionnaire

1. Until now, have you believed Darwin's theory of evolution to be true? If *'yes,'* on what evidence, broadly, did you rely? If *'no,'* what caused you to *doubt?* Are you well read on this topic, or have your ideas been driven more by the general *cultural assumptions* of our era, where evolution, as Darwin described it, is widely felt to be just another **fact of Life**, like gravity?

2. If it turns out that Darwin was wrong will you feel surprised or let down? Or was it just that he and his followers saw what they *wanted* to see and believe? For example, they dropped a too literalistic religious creationism, which tries to treat religious creation stories as if they are literally true, only to replace it with Darwin's *'Just So Story,'*[3] which, unlike religious creation stories, was ***supposed*** to be *scientifically* **accurate**. It was not meant to to be merely an alternative creation metaphor for materialists and atheists.

3. Personally, I think that *Life, and Mind, either* have intelligent causes, which, mysteriously and <u>impossibly</u>, *'just are,' or,* they have MINDLESS causes which mysteriously and <u>impossibly</u>, *'just are.'* Does this binary make sense to you?

4. Why '<u>impossibly</u>?' Because *how* can there be anything-at-all, including, most mysteriously of all, an ultimate *Cause of All,* rather than nothing, and no-cause-at-all? *What* could cause that *Cause* to be, and *its* cause before *that?*

The same questions repeating forever. Or, do we simply acknowledge that, *logically,* there must be a *Final or Uncaused Cause,* of everything else, which, *'impossible'* or not, *'just Is'?* Does this line of thinking make sense?

5. If we can never know *Why* there is a *Cause of Anything* at all, rather *nothing* at all, can we ask which—of the two possible causes of anything at all, versus nothing nothing at all—that is, MINDLESS *vs.* **intelligent**—is the more consistent with the data, and the more logical overall?

Do you feel a greater intellectual pull to one or other of these two basic options—**intelligent** causation vs MINDLESS? If so, why? Do you feel an opposing emotional pull? If so, why?

6. If, like Darwin, we incline to materialism, in the absence of *Creative Spiritual Forces,* we *have* to believe that *pure chance* is a highly creative force. What, though, if you have a view on this, is the *scientific* evidence, if any, for the *(allegedly)* amazing creative powers of *pure chance?* To create cells and shells, blood and brains, eyes and minds—to give rise to *feeling* and *thinking, love* and *joy,* wonderful *curiosity* and intelligent *creativity?*

7. Some say that modern science has shown us that belief in the *super-natural* is just wishful thinking. Is this true? Or, is this just **materialism** dressed up as science? Whereas, science is not materialism per se, but a range of methods for investigating **all of reality,** be it purely PHYSICAL, or also *spiritual?*

If the *super* natural *is* real, are there any reasons for us ***not*** to *investigate* it?, where possible?, and as much of it as possible?, using logical, modern, scientific means? Is this a good idea? Or, is it wrong in some way? If so, why?

8. When it comes to origins, the position in science today is (or seems to be) that scientists are clever enough to work out that *Existence* is, fundamentally, MINDLESS at its core, and in its causation, but *not* clever enough to work out that this may *not* be true—to discover that, actually, *Nature's Laws, Cells and Shells, Stars and Planets, Life and Mind,* may be but the outward expressions of an incomprehensibly powerful, inner, *Spiritual Intelligence.*

Scientists are encouraged to argue for "unintelligent design" or UD, and for the materialist world view, as Darwin and Dawkins do, but if they argue for "intelligent design" or ID, using modern scientific means, as various more modern thinkers, like Behe, Dembski and Meyer do, we are to doubt their abilities to give us any good evidence for *design and intelligence* in Nature. *Should* we doubt their abilities? Or does this needlessly *restrict* our scientific inquiries and, potentially, lead to a far too narrow view of holistic reality?

9. Is science just a matter of consensus and majority views holding sway? Or do *minority* views and *alternative* hypotheses also have a role to play?

10. Is it ok to let go of an old theory, like Darwin's, and to follow *newer* evidence, and better thinking, which *falsifies* the old theory? Or, is it best to stick with the old theory, not because it's true, but because the alternatives are currently unfashionable, emotionally difficult, or politically inconvenient?

11. Can we take anything of value from an unknown author like this one? Or should we always just follow the authorities, the schools and the universities, *Wikipedia,* and the MSM, the fashions, and the majority views of the day?

1 The Trouble With Creation

'Do you believe in creation, Alisha?'

'Ah, that's a *big* question! But, if you mean: 'Can we read religious books as if they contain scientific descriptions of *Life's* origins?' No, I don't. If you mean, "Do I think there is *more* going on with nature, especially with *Living* things, than the blind chance of modern atheist belief?" Then, yes. Whether we call the *incredibly intricate* processes, which create and sustain life, 24/7, *creation* or something else doesn't matter to me. What about you?'

'Well, I think we have always tried to make sense of our origins, and, at the simplest, there are just two ideas: either *Life* arises intelligently, or, it is created by pure chance. Either the universe has *intelligent* causes which, very mysteriously, and <u>seemingly impossibly</u>, *'just Are,'* or it has MINDLESS causes which, just as mysteriously, and just as impossibly, *'just Are.'*

'Why *'impossibly'?'*

'Because, *how* is it possible for there to be *any.thing* at all, including a *Cause of any.thing at all,* rather than *no.thing* at all? Isn't it easier to imagine there being no-thing rather than some-thing? Yet, *Being, and Beings, beautiful Existences,* like amoebas and bees, flowers and we, clearly *Are,* and the seemingly impossible has happened.

'Ok. But we can, presumably, try, Ollie, to work out whether the ultimate *Source of Being,* which gives rise to *all Beingnesses,* flowers and trees, rivers and seas, and curious, creative Beings, like you and me, is an amazing intelligence which *just Is,* or a totally MINDLESS mechanism which *just is?*

'Yes, I think we *can* work some of that out. For example, for a theory of creation by pure chance to be true, we must assume that apparently mindless atoms like CARBON and HYDROGEN must, somehow or other, have luckily bumped and clumped together, and, after a very long and very lucky evolution, eventually become people-shaped, able to *think,* and to ask questions like: *"Who* are we? *Where* do we *come* from? Is there any **meaning** to our lives?" If, on the other hand, creation by *spiritual intelligence(s)* took place, *how* did it do so? Was it very slowly, over vast time, or was it rapidly, and, as some say, even as recently as a mere 6000 years ago.'

'You the mean the bible-based idea known as *young–earth* creationism?'

'Yes. Others, less literal minded, argue for *old–earth* creation, where some elements of reality were created more or less instantly, like the Big Bang, and others were evolved, over time, perhaps along Darwinian lines, but with some allowances for *super* natural *intelligence* and design.'

'But neither creationism nor intelligent design are considered intellectually reputable today, are they, Ollie? I'm not a *big* fan of organized religion, but I disagree with many of Darwin's followers who deny the existence of the *spiritual* altogether. They insist that Life is just a cosmic accident, which evolved by accident, that Darwin *proved* it, and that it is naive of me to believe in anything more than PHYSICAL REALITY.'

'Ok. Well, unlike you, Ali, I did go through a period, in my teens, when I did very much believe in Darwin, and I became very *atheistic* in my thinking, like a young Richard Dawkins. It was only later that I worked out that there was so much that Darwin's ideas—and the materialist philosophy which was driving them—*did* not, and *could* not, make sense of.

'But, the idea of 'creation' by pure chance is very popular today, and Darwin's version of this idea is the most famous. He, is, in fact, regarded by many as the materialist (atheist) school's greatest hero. Forget *God, or Spirit,* we do not need these to explain anything. "Life is just a long series of chemical accidents, which evolved by accident, and Darwin proved it."

'In fact, Darwin's ideas are, now, so mainstream that to express any doubts about them is to risk ridicule and to being asked, in a tone of condescending disbelief, "You're not some weirdo *creationist* are you?"

'But why reject all concepts of *intelligent* causation, even if we leave off young earth creationism, which is so doubtful, scientifically?'

'I think it's because, by the middle of the 19th century, many intellectuals had become less and less willing to be told what to think by the traditional religious authorities, who had, so often, been so dogmatic and oppressive. It was also because a medieval bishop called Ussher's idea that the Bible could be read like an early science text, which tells us the precise age of our planet, began to make less and less sense. Modern geological discoveries had shown earth to be *far* older than the 6000 years he allowed for, based on his very careful, but far too literalistic, analysis of the timelines in the bible.

'This is why many dropped creationism altogether, and opted for Darwin. He, they felt, had supplied a totally **naturalistic** explanation of Life's history, so the *'spiritual'* was no longer needed to explain anything. He had found a '**god-not-needed**' explanation for Life and 'logic' and 'reason' had, finally, triumphed over 'blind' religious belief and superstition.

'I guess, Ollie, if we have very childlike ideas of the creative spiritual forces, looking, for example, like an ancient patriarch, with a long flowing beard, a bit like Father Christmas, then we either (a) continue with those, or (b) we understand that such stories are *symbolic* expressions of spiritual truths, or, lastly, (c) we question the existence of anything *spiritual* at all?'

'Which, Ali, is pretty much what happened, in the West, especially in the 18th century, in the time known as the European Enlightenment. Even more so after Darwin published his ideas on evolution in 1859. But, as you say, what many of us do not notice, is that it is our *own,* straw man ideas of creation that are the problem, not the classical idea of *intelligent causation* itself, versus the totally mindless 'creative' processes Darwin championed.

'As 'right-thinking,' modern, 'intellectuals,' we set up laughable, straw man Father Christmas type gods, to represent the Ultimate Creative Force or Forces, which no serious *spiritual* believer believes in, or *needs* to believe in, then we delight in making fun of these, pointing out that they're absurd, before setting up our own, even more absurd pseudo god in their place.

'You mean the materialist 'god' of Pure Mindless Chance or PMC?'

'Yes. It is strange, Ali, but most of my atheistic friends do seem to find it hard to understand that *Spirit* looks nothing like Father Christmas! But, rather, *God, Spirit, the Tao, Supreme Intelligent Beingness,* (or whatever label we prefer), is more like an *intrinsic, all-pervading, Blazingly Intelligent, Living, Loving, Light, Life-Force Energy,* not a bearded-god type figure!

'A *brilliantly, stupendously, intelligent energy* which expresses creation in many different ways, not just on our planet, but throughout the vastnesses of the stars and galaxies, in endless variations, exploring endless possibilities of **outer forms** and *inner sentiencies,* for the myriad different kinds of

27

beings, PHYSICAL, *and non-physical,* including the *angelic* and other *spiritual hierarchies…'*

'A creation, which, by the way, Ollie, isn't just in the past, but is happening *this* very moment, and *every* moment, *second!, by second!, by second!,* keeping my heart beating, and yours, your lungs breathing and mine.

'We take all these *amazing things* so for granted, never stopping to think of the truly staggering levels of intelligence involved!'

'I agree. Darwin's once novel idea that Life is *not* intelligently created, but is just a long running series of chemical accidents seems absurd to me. But his strangely unrealistic views are mainstream today. What will it take, though, for us to wake up from our strange modern *faith* in Darwin's 'god' of pure mindless chance or PMC? Not that it is easy for us to wake up in this area, as so many of our teachers, scientists, academics, and the mainstream media (MSM) promote these strange ideas: that science has shown us that:

(a) Reality is **wholly PHYSICAL**—there are no *spiritual* dimensions.

(b) The universe started with a random Big Bang which *just was*.

(c) Following this random Big Bang, which *just was,* there arose physical laws of such elegant functionality, and cosmic fine tuning of such amazing precision, that the best explanation for it all—apart from the classical, *God or Higher Power* hypothesis—is MULTIVERSE THEORY.'

'What is multiverse theory? It sounds intriguing . . .'

'It is the proposal that there *may* be multiple universes, perhaps forming all the time, due to a mysterious but totally **Mindless, Unintelligent, and Accidental, Universe Generating Mechanism** or MUA-UGM which *just Is*—for no one can say what caused the MUA-UGM to so mysteriously *just Be*—and eventually, this stupendous, but totally mindless mechanism, which *just Is,* 'might' give rise to a universe which looks and works just like the universe we find ourselves in. After all, if the MUA-UGM is generating millions of universes per day, or every second even, it may, at some point, get lucky.

'That sounds absurd, Ollie. Perhaps the 'multi-verse' is just a way to avoid the classical idea that the universe we are *actually* in, is intelligent and meaningful, not utterly mindless and meaningless. The idea that *Life—and*

Mind—senses and feelings—breathing and eating—hearts and lungs—blood and brains, could arise in such a mindless way seems absurd to me . . .'

'I agree. The classical hypothesis of *spiritual* causation makes much better sense of the data, overall. But the *mindless* multiverse is a fashionable view today. Because, as you say, its fans think it avoids the idea of *spiritual* creation, which they dislike. Although, ironically, the *mindless* multiverse hypothesis is, I think, a far *less* rational idea than the classical *intelligent* universe idea. John Lennox, an Oxford math professor, favors the classical view.

'He draws our attention to the astonishing **cosmic fine tuning**. He writes,

> 'Theoretical physicist **Paul Davies** tells us that, if the ratio of the nuclear strong force to the electromagnetic force had been different by 1 part in 10^{16} [1/10,000,000,000,000,000], no stars could have formed. Again, the ratio of the electromagnetic force-constant to the gravitational force-constant must be equally delicately balanced. **Increase it by only one part in 10^{40}** [10 with 40 zeros after it] and only small stars can exist; decrease it by the same amount and there will only be large stars. **You must have both large and small stars in the universe**: the large ones produce elements in their thermonuclear furnaces; and it is only the small ones that burn long enough to sustain a planet with life.' [4]

'This is a *staggeringly* precise condition, Alisha, but how can those of us, who are not astro-physicists, even begin to grasp just how stupefyingly exact it is? Oxford mathethematics professor, John Lennox's account of this illustration, suggested by the astrophysicist, Hugh Ross,[5] is instructive.

> 'Cover America with coins in a column reaching to the moon (236,000 miles away), then do the same for a BILLION **other continents of the same size**. [Then] Paint ONE COIN RED and put it somewhere in ONE of the billion piles. Blindfold a friend and ask her to pick it out. The **odds are about 1 in 10^{40}** that she will.'

'As if this is not astonishing enough, Lennox mentions that the eminent cosmologist Sir Roger Penrose calculates that in order to get to the kind of universe we are actually in and to one which will actually work, 'the 'Creator's aim' must have been accurate to **1 part in 10 to the power 10^{123}, that is 1 followed by 10^{123} zeros**, a 'number which it would be impossible to write out in the usual decimal way, because **even if you were able to put a zero on every particle in the universe** there would not even be enough particles to do the job.' [6]

'Faced with not one, but many such spectacular examples of fine-tuning, it is perhaps not surprising that Paul Davies says, '**The impression of design is overwhelming**.' [7]

'So far, though, this amazing cosmic fine-tuning data does not disturb a now almost automatic belief in blind-chance explanations for everything.'

'Why, Ollie? It doesn't make sense. After all, is there *any evidence,* at all, from *any* time or place at all, that mere chance, as opposed to the classical idea of *spiritual intelligence,* can create anything remotely clever?

'Not as far I'm aware, Ali. It's just a strange modern *faith*. Yet, due to this peculiar modern faith in pure chance, as a supreme creative force, there is, nowadays, little to no discussion as to whether this actually works, nor whether there may be any *spiritually causal* components to reality.'

'But don't some of Darwin's followers say, "Couldn't there be a *spiritual* creation, but one which intelligently evolves, rather than instant?" Mused Ali.

'Yes, and it's not a bad idea. It is what some old-earth creationists and esoteric thought schools like yoga, theosophy, and anthroposophy think.'

'But it was not Darwin's idea? Evolution was not *intelligent?*'

'No. Because a *key* part of his thinking was his idea that, maybe, there is **no** intelligence or purpose to *Living Nature*. No *God, no Spirit,* just **mindless matter.** Perhaps, there were just 'dumb' matter + 'dumb' laws, generated by a Mindless Unintelligent and Accidental Universe Generating Mechanism or MUA-UGM, which *just Is,* + millions of lucky chemical coincidences, and these led to all things, including to you and me.

'The mysterious MUA-UGM, which *just Is*, materialism's substitute *Final Source* and *Uncaused Cause,* which is not so different from the classical *God or Spirit* of conventional spiritual belief, except that it is mindless, it has no *inner Nature, no Subjectivity, no Soul, no Desire, no Creativity, no Love, no Intelligence, nor Purpose, nor Will* of any kind . . .'

'But does anyone *really* believe, Ollie, including those famous atheists, whose names we know, that such a mindless mechanism could *really* generate the kinds of living, loving, feeling, creating Beings we find ourselves to be? Isn't it totally bizarre to look at another Human being, a soaring Eagle or a beautiful Butterfly, and to maintain that, in essence, all we are observing is just a chemical coincidence—because Darwin said so?'

'I agree, it is very odd. Not so long ago, the answer to your question would have been a universal, *'Yes!'* The classical view being that *Spiritual Mind* came before [apparently] S..O..L..I..D   M...A...T...T...E...R[8] and the cosmos was the vasty space in which *Spiritual Ideas* were made visible.[9]

'A theory of evolution which reflected this understanding, would allow that there may be *more* to Life than the physical alone. Whereas, a totally materialistic or atheistic theory of life, like Darwin's, which, in our time, Dawkins so promotes, not only **restricts scientific inquiry**, because it rules out *spiritual* causation on principle, but it asks us to put a truly *irrational* level of *faith* into mere chance as a supreme creative principle.

'So why do so many people, now, go along with this unrealistic idea?'

'Because Life's deeper sources cannot be *physically seen?* They can only be *inferred,* or known, to some extent, by *inner, spiritual* means.

'Because, so many today dislike religion?, which they blame for many of the world's troubles. This is why well known, modern fans of Darwin, like Dawkins, publish book after book in which they urge us to believe that,

> "Dear public, Life is nothing miraculous, in the old sense of the term. There is no *God, or Spirit, or Creation.* That's wishful thinking. Life is a purely 'NATURAL' phenomenon, like the wind and the rain. Although *living things* are very complex, and they do *look* designed, as even Darwin admitted—think of the magical Butterfly or the Eagle's amazing eye, think of the Peacock's lovely tail, or the incredible Human mind—actually it all arose, and it evolved from there, from MINDLESS elements, like CARBON, HYDROGEN and OXYGEN by pure, mindless, chance (PMC).
>
> "As Dawkins says, "In a universe of…blind physical forces… some people are going to get hurt, other people are going to get

lucky, and you won't find any rhyme or reason in it, nor any justice. The universe… has precisely the properties we should expect if there is, at bottom, **no design, no purpose, no evil, no good, nothing but pitiless indifference.**" [10]

"Sorry to disappoint you folks, *but, the science says,* this is just how it is. Darwin, possibly the greatest scientist of all time, showed us that the mindless laws of chemistry, *plus* enough time, *plus* enough lucky chemical coincidences could, and did, create all."

'In ***Heretic: One Scientist's Journey from Darwin to Design***, Matti Leisola, quotes a scientist colleague, Esko Valtaoja, who once said to him,

"Life is nothing else than physics and chemistry—[it's] mere electricity. There is **no reason** to assume anything supernatural." [11]

'Valtaoja expresses what, after Darwin published, became the mainstream scientific view, as far as a *spiritual or intelligent creation* is concerned: there *isn't* any. "There is no reason to assume anything supernatural," he so confidently tells us—or miraculous—or spiritual. *God is a Delusion,* as Dawkins has famously written.'

'And anyone who disagrees is uneducated or emotionally fragile?'

'Yes, that tends to be the attitude in the mainstream science community, today, especially here in the intellectual west. Here, where **93%** of the members of the National Academy of Sciences, the most prestigious science body in the USA, profess not to believe in *God* or the *Spiritual*.[12]

'Here, where, especially in science, we have become ever more materialistic in our thinking—incapable, it seems, of conceiving of anything beyond that which we can PHYSICALLY see, hear, grasp, measure, or control.

'Nothing *inner,* nothing *spiritual,* nothing *cosmically* intelligent, nothing truly *soulful* can be allowed any place in this depressing, mono-dimensional *ontology* of meaninglessness—which various well known thinkers promote, year after year, after year. No soul, no spirit, no meaning, no point.'

We continued on our way.

2 Cosmic Accident

'Ironically, it was at the dawn of the modern age that some thinkers tried to approach the scriptures themselves in a more *scientific* way. Like, Bishop Ussher's careful, 17th century analysis of the time lines in the Bible, which led him to conclude that Earth was only 6000 years old. This is where today's young-earth creationists get their dates from. However, in the 18th Century, not long after his analysis, geologists realized earth's far more ancient age. Their discoveries falsified Ussher's young-earth time lines.

'But discovering that earth was much older than Ussher thought, did not disprove the *spiritual* . . .' Mused Ali. 'Falsifying young-earth creationism does not falsify *intelligent causation,* as such, does it?'

'No, because people could, still, hypothesize old-earth creation.'

'But I guess that might not have satisfied those who had, by then, begun to look for something more *scientific* than: "Just read the Bible…"

'Ok, that's true, in the sense that this was a time when scientists like Darwin's grandfather, Erasmus Darwin, and the French biologist, Jean-Baptiste Lamarck, began to propose their own, early theories of evolution, *vs* traditional ideas of creation—as a more or less *instantaneous* event.'

'So evolution was in the air some time before Darwin published?'

'Yes, but, *unlike* his grandson, Charles Darwin, and most of Charles' later followers, Erasmus Darwin allowed for some *super* natural input. Perhaps a remote, *deistic god,* 'THE GREAT FIRST CAUSE,' had set up the conditions for life to automatically evolve from there. What do you think?'

'Well, unlike famous modern materialists like Dawkins, I don't think we can leave the *super* natural entirely out of our ideas, not, that is, if we want to make the *best sense* of Life on Earth. But, at the same time, I *can* understand why some people, like Darwin and his grandfather, Erasmus, wanted to make more *scientific* sense of Life's origins, than simply relying on ancient religious texts, which, although they may contain many valid *spiritual* truths, cannot be treated as science.'

'Of course, and I agree. My difficulty is with the fact that, after Darwin published, scientists increasingly assumed that either the *spiritual* did not exist at all, or that, *even if* it did, there was no need to refer to it in our scientific explanations of the 'natural' worlds. Even the very *Miracles of Life, of Birth, of Mind, of Feelings,* all came to be considered to be purely PHYSICAL phenomena, no more *essentially* mysterious than the wind and the rain, which we can explain well enough in purely PHYSICAL terms.

'This is why, since Darwin published his atheistic theory of evolution, existence has, in western scientific circles, come to be seen, more and more, as a mere cosmic accident. The very fact *of Existence, of BEING,* vs *non-being,* no longer a *miracle.* God, as Nietzsche famously said, was dead.'

'But that's so sad, Ollie. It makes existence seem pointless, and it means that there's no continuity of *Life* after this life. Death is truly a full stop.'

'Although, what if that gloomy view is correct?'

'You mean that when the chemicals stop we drop . . . and that's it?'

'Yes, because as truth seekers, we must care for what's *true,* not just what we hope is. So, if Darwin was right, we must accept it and move on. On the other hand, if he was wrong, and if it can be *shown* that he was wrong, it will free science up for newer, fresher inquiries into origins. Then, Life will become so much broader and more interesting to explore once more.'

'But are there any ways to show that Darwin was mistaken, if he was?'

'Yes, there are, many actually. Although the idea that, "Life is just an accident, which evolved by accident," was not, despite his most famous work being called *'**On the Origin of Species*** a formal part of Darwin's theory.'

[Note, the full title: *"On the Origin of Species by Means of Natural Selection, or the **Preservation of Favored Races** in the Struggle for Life."* Unfortunately, Darwin confirmed some peoples racial prejudices. The Nazis liked Darwin. His ideas had many consequences.]

'So Darwin's title was not true? That's strange . . . '

'Well, despite the claims, Darwin did *not,* in honest fact, explain the origin of the species *at all.* He just *assumed* that Life *might* have started by chance—as a random, single, microbial cell, like a bacterium—which *might* have *varied,* and *varied,* until, by two hypothetical processes, which Darwin called (i) **natural selection** and (ii) **descent with modification**, it might, eventually, have become all today's species. An idea which, as we'll see, is not only unlikely, **and unproven to this day**, but demonstrably *impossible*. Yet, unrealistic as it was, it did not stop Darwin writing to his friend Hooker,

> 'But if *(and oh **what a big if**)* we could conceive in some warm little pond with all sorts of ammonia and phosphoric salts – light, heat, electricity etc present, that a protein compound was chemically formed, ready to undergo still more complex changes'[13] [and, by pure chemical luck, become Life, in the form of a *Living Cell,* 'the most complex system known to man.'[14]]

'Then, what started out, in Darwin's private letter to his friend, as a *very hypothetical **'if,'*** has, over time, morphed into one of the *stranger dogmas* of modern science—despite the **total lack of any evidential support** for it.'

'That *Life* arose by pure chance, as he suggested to his friend?'

'Yes. Many do now believe that. But, for an idea to be *scientific* it must be **testable,** otherwise it's just theory. So, let's test the claim that:

> "Darwin and other scientists have shown us that, if we allow enough time, *anything,* no matter how complex, can be created by pure chance. So, there's no need to refer to the *super* natural, to anything *beyond* outer, physical NATURE to explain all of Life."

'In Darwin's day, it was thought that the kind of life which might play the role of his hypothetical microbial ancestor-to-all-later-life, was relatively simple. So, Darwin wondered whether cells might not arise ***naturally?***

'So he believed cells were fairly simple mechanisms?'

'Yes. That is what people then thought. Darwin's contemporary, Haeckel, called the cell a 'homogeneous globule of protoplasm.' So, why shouldn't they arise naturally? Darwin and Haeckel were, however, very mistaken.'

'You see, Ali, cells, they did not then realize, are astonishingly complex. It is easy to demonstrate that they could *never* arise by mere chemical luck.

'Although tiny, each cell is **vastly more complex,** at its own level, than the much larger organs, like lungs, livers, or kidneys, it helps to build.

'For example: We can, nowadays, build **heart or kidney machines.** For us, to drill rocks and oils from the ground, and to turn them into these amazing machines, is a *very clever* thing to do. But, the idea that we could

35

build even one, *dynamic,* **Living***, self-reproducing,* **self-repairing** *cell, from scratch,* one tiny cell, which, internally, is *far more* complex than any heart, lung, or kidney machine, or, indeed, anything we make, would be laughable.

'Every cell in our bodies, each smaller than a sand grain, is, at its own level, like a vast, micro-miniaturized, super-computer-controlled chemical factory, *far more complex* than any of our factories, or, even whole groups of them. Cells are stunningly—mind-blowingly—complex.'

'So there is **no evidence** for the idea that Darwin put to his friend?'

'No, there is not the tiniest bit of data to say that mere luck can create any Living thing. Nor, that it can give rise to *two key* features which distinguish all living things, and, indeed, all the mechanisms we make. Namely, that, all living things are both *complex* and **specific**.'

'Go on...'

'Well, not only are the *Life* forms very *complicated,* that complexity is not random. Imagine a Car, TV, or PC. It must be assembled in a very **specific** (or non-rA*n*do*m*) order or it won't work. The point being that mere chance *cannot* give rise to this kind of **non-random** complexity. The laws of probability and the laws of entropy forbid it. It should be obvious.'

'You mean, if we *r*anDo*m*L*y* tumbled Car, PC or TV parts around, for a long time, we'd never get a working machine?'

'Exactly. The mix would be *complex*, but non-functional—no functionally arranged sequences of parts. Any 'natural' ordering would, at best, cause the heaviest parts to sink to the bottom of the mix.

'As contemporary theorists of intelligent design, or ID, like Stephen Myer and William Dembski, have pointed out, REGULAR NATURAL laws, on their own, do not have the necessary creative power, nor any reason, to generate the totally **specific sequences** of parts and software which characterize all computer CODES, all machines, and all Living things.

'But isn't it **utterly obvious** that random interactions will *never* give rise to any *complex* mechanism? Why do we need to argue for this?'

'I totally agree. But, as a culture, we seem to have forgotten. We pride ourselves on being *'scientific,'* but the idea that *unintelligent* chance can create anything remotely clever is about as unscientific, *and non-rational,* as it is possible to be! Unfortunately, when it comes to origins, we are in a weird, *Emperor's New Clothes* situation where those brave souls who point out—what should be the blazingly obvious—that mere chance does *not* have the vast creative powers claimed for it, are ignored or laughed at.

'What we *do* know is that chemical coincidences in nature can give rise either to rAnd0m c0mPlexity, as in a tornado or wild fire, or to SIMPLE, SPECIFIC patterns, as in crystal formation. Patterns which can be described by simple algorithms like ABC + ABC + ABC. Neither of which are capable of generating Life. As leading origins of life researcher Lesley Orgel writes,

> 'Living things are distinguished by their **specified** [specific] complexity. **Crystals** such as granite [although **specific**, consist in simple patterns so they] fail to qualify as living because they *lack complexity;* mixtures of **rAnD0m** polymers [although they are *complex*] fail to qualify because they *lack* **specificity**.' [15]

'A regular physical law which can generate simple **specific** sequences, like ABC + ABC + ABC, cannot generate the far more complex sequences required by instructions like TURN + THE + SWITCH + ON + HERE.

'A sequence like this is *physically* indeterminate. Due to its **specific** but *irregular* content it cannot be determined by physical laws alone. Such a sentence can only be created by intelligent design—whatever its source.

'Similarly, a specific, DNA CODED sentence like ATAGGAGC + CAACGGTT + CCTTGTAC cannot be determined by natural laws alone, because it is **complex,** it is **specific**, and it is iRregUlar. No CODE or software can be produced (i) by chance alone (ii) REGULAR laws alone or (iii) a mix of the two. The laws of probability forbid it. It should be obvious.'

'Ok, but what about the spark and soup experiments of the 1950s? Some of my friends say that Miller and Urey showed how Life *could* have started by pure chance, that *Life* is just a 'natural' phenomenon like any other.

'Some of mine too, but it's a false claim. All the experiments generated were a few amino acids. But a few amino acids don't make a Life Form, do they? Such chemicals unguided by *higher-level* forces, will never lead to anything but samples of acids. As biologist Michael Denton notes:

> 'Instead of revealing a multitude of transitional forms through which **the evolution of the cell** might have occurred, molecular biology has served only to emphasize the enormity of the **G A P**.
>
> 'We now know not only of the existence of a **break** between the **living** and **nonliving world**, but also that it **represents the most dramatic** and fundamental of all the **discontinuities** of nature.
>
> 'Between a *Living Cell* and the most highly ordered non-biological system, such as a crystal or snowflake, there is a C H A S M as vast and absolute as it is possible to conceive.'[16]

'We need to let that sink in. Yet, despite this, we are, currently, urged to believe that this CHASM, between **NON-LIVING** and *Living* things, *could* be bridged, *was* bridged, by *pure chance,* and to go along with this quite extraordinary claim. **Carl Sagan**, a well known 20th century intellectual, said, skeptically, of some people's religious and spiritual beliefs:

"**Extraordinary claims** need to be backed up by *extraordinary evidence."* Fair enough. Yet there is no support, *at all,* for our current science culture's quite extraordinary beliefs in *pure chance* to create anything remotely functional—**DNA CODES**, cells and feathers, blood and brains, snails and whales, fish and birds, monkeys and people. Again, despite this, we have often heard the utterly unscientific urban legend, with no supporting evidence, of any kind, let alone any **extraordinary** evidence, that:

> 'Just give it enough *time,* and a few randomly typing monkeys, would, eventually, produce the complete works of Shakespeare. Such is the power of blind-chance causation! It's just a question of **time** and **probability.** Pure luck can, in theory, build anything: **CODES**, AI guided Robots, Living Cells of stunning complexity, even entire Animals, or Humans, orders of magnitude more complex than any computer **CODE** or robot we make.'

'Ha ha, Ollie. But surely this modern belief in pure chance, as a supreme creative agency, is a non-starter for explaining Life's origins?'

'Yes, of course it is. It is a total non-starter . . .'

'So why do people take it seriously? Why is it so fashionable today?'

'Because, life's deeper sources are not obvious. Because many of us now reject *spirit* out of hand. Because, scientists, today, are, literally, trained to assume that just **chance + necessity** (NATURAL laws) can give rise to anything.

'The trouble is, the creative capabilities of pure luck have been grossly overestimated. Consider: in his book, *Genesis and the Big Bang,* physicist Gerald Schroeder looked at the odds of monkeys randomly typing out just one Shakespeare poem, *Shall I Compare You to a Summer's Day?'*

> 'There are 488 letters in the sonnet . . . The chance of randomly typing the 488 letters to produce this one sonnet is one in 26 to the 488th power, or one in 10 to the 690th power. The number 10 to 690 is a one followed by 690 zero's! . . . [Yet] since the Big Bang, 15 billion years ago, there have been only 10 to the 18th power number of seconds . . .
>
> [In fact, at] one random try per second, with EVEN A SIMPLE SENTENCE having ONLY 16 LETTERS, it would take 2 million, billion years (the universe has [only] existed for about 15 billion years) to exhaust all possible combinations.' [17] [We need to *think!*]

'An article in wikipedia, *Infinite Monkey Theorem,* makes a similar point:

> 'Even if *every* proton in the observable universe were a monkey with a typewriter, typing from the Big Bang until the end of the universe ..., they would still need a ridiculously longer time - more than three hundred and sixty thousand orders of magnitude longer – to have even a 1 in 10^{500} chance of success[fully typing out all of Hamlet]. . . .
>
> 'In fact there is *less* than a one in a trillion **chance** of success that such a universe made [entirely] of monkeys could type any particular document A MERE 79 CHARACTERS LONG.' [18]

'When it comes to living things, William Dembski, a modern intelligent design (ID) thinker, shows us how *Life,* also, could **never** arise by chance. Dembski agrees that a mechanism's *non-random* patterning, its **specificity** should be extremely high before 'mere chance' can be ruled out as a possible cause. But, he explains, at a certain point, a particular sequence of information, or of parts, in the case of *Living Cells,* polypeptide chains,

> "becomes [just] too long, [just] too improbable, [and] too complex to reasonably attribute to unspecified, unintelligent chance." [19]

'Regarding Darwin's chemicals in a 'warm little pond' idea, Dembski explains that—never mind a little pond—even if the *entire* universe had consisted in a pre-life chemical soup, even if it had been *deliberately* set to generate random amino acid sequences, by using every last particle in it, even if it had been *doing nothing* else since the Big Bang, it would *never* lead, by mere chance, to even just one of the *thousands* of specific protein sequences needed to make Life possible.'

'Because…'

'Because the laws of probability forbid it. It is demonstrably *impossible*. However, even if any one of the *thousands* of needed sequences did arise, by chance, with **no cellular membrane** to protect it, it would be as rapidly destroyed by the other chemical-reactions going on around it, 24/7.

'Remember, it's the DNA *inside* the Cell which codes for the same, very clever—selectively permeable—cell wall which protects the DNA *inside*.

'There is no way the DNA *within* could arise, by *chance,* to CODE for the very membrane which protects the DNA from *without*. These two factors are **irreducibly complex**. They *must* arise *together,* or not at all. Yet this fascinating—and perplexing—chicken or egg issue is ignored.'

'In their book, *Intelligent Design Uncensored,* Dembski and Witt show how scientists today are able to use **scientific** methods detect evidences for intelligence and design in nature *vs.* mere chance or 'unintelligent design.'

'So, they ask: is the arrangement of the parts of a basic cell complex enough to pass a rigorous—*scientific*—test for detecting intelligent causation in nature, versus the kind of only-seeming design which can account for pretty wave-made patterns on a sandy beach. They write,

> "How improbable does a **specified** thing [such as a **specific** sequence of CODE or parts] have to be before we can know it was

designed? One chance in ten is obviously too low a threshold. One chance in a hundred is too low. How high is high enough?

"For our purposes, ... we want to set the bar very high, meaning the thing in question will have to be *extremely* improbable to pass our design test." 20

'In ***The Design Inference***, Dembski sets the threshold for detecting real design—vs. mere chance—very high, at 1 in 10,^{150} 1 followed by 150 zeros. This is because there are about 10^{80} elementary particles in the universe and we now know that matter can change from one state to another no faster than 10^{45} or 10,000,000,000,000,000,000,000,000,000, 000,000,000,000,000,000 per second.

'So, if we suppose, he writes, "that any event in the universe requires the transition of at least one elementary particle... then the total number of events throughout cosmic history [14 billion years] could not have exceeded 10^{80}x10^{45}x10^{25} => 10^{150}." 21

'So, he asks, "Is the arrangement of the parts of the most basic cell,"22 complex enough to pass such a test for *real* design in Nature—versus the kinds of chance-based processes, which Darwin imagined with his **'*big if,*'** chemicals-in-a-warm-little-pond idea? Dembski continues:

"Scientists calculate that a cell with just enough parts to function in even a crude way would contain at least 250 genes and their corresponding proteins. The *odds* of the early Earth's chemical soup randomly burping up such a micro-miniaturized factory are unimaginably longer than 1 chance in 10^{150}." 23

'So, just as with the sequences of letters in Shakespeare's poetry, living things could *never* arise by chance?

'Yes, it is not merely unlikely, **it is impossible**. Michael Denton writes:

'Although the tiniest bacterial cells are incredibly small . . . each is in effect a veritable **micro-miniaturized** factory containing thousands of exquisitely designed pieces of intricate molecular machinery, made up altogether of one hundred thousand million atoms, **far more complicated than any machinery built by man** and absolutely without parallel in the non-living world.' [24]

'Darwin and his colleagues had no idea that, as Denton outlines, in his influential, 1985 book, ***Evolution a Theory in Crisis,***

'To grasp the reality of life as it has been revealed by molecular biology, we must magnify a cell 1000 million times until it is **20 km in diameter and represents a *giant airship* large enough to cover a great city like London or New York** . . . On the surface of the cell we would see millions of openings, like the portholes of a **vast spaceship,** opening and closing to allow a continual stream of materials to flow in and out. If we were to enter one of these openings we would find ourselves in a **world of supreme technology and bewildering complexity**. We would see endless highly organized corridors and conduits branching in every direction away from the perimeter of the cell, some leading to the central memory bank in the nucleus and others to assembly plants and processing units. **The nucleus itself would be a vast spherical chamber more than a kilometer in diameter**, resembling a geodesic dome inside of which we would see, all neatly stacked together in ordered arrays, the miles of coiled chains of the DNA molecules. **A huge range of products and raw materials would shuttle along all the manifold conduits** in a highly ordered fashion to and from all the various assembly plants in the outer regions of the cell.

'We would wonder at the level of control implicit in the movement of so many objects down so many seemingly endless conduits, all in perfect unison. **We would see all around us, in every direction we looked, all sorts of robot-like machines** ... We would see that **nearly every feature of our own advanced machines** had its analogue in the cell: artificial languages and their decoding **systems, memory banks for information storage and retrieval**, elegant control systems regulating the automated assembly of components, **error fail-safe and proof-reading devices used for quality control, assembly processes** involving the principle of prefabrication and modular construction . . . However, it would be a factory which would have one capacity not equaled in any of our own most advanced machines, for **it would be capable of replicating its entire structure within a matter of a few hours** ...' [25]

'Do you really think, Ali, if Darwin and his friends had known of these things, they would *not* have revised their views? They would, surely, have thought it very strange to maintain, in the light of all our knowledge today, that such sophisticated processes might arise by pure "luck," with no more organization needed for them than the kinds of processes involved in **basic physics or geology**—as Darwin had imagined might be possible. They had no idea how **just how complex** living things would, eventually, turn out to be.

'Unfortunately, once certain 19th century thinkers realized that the old religious creation stories were not *literal* descriptions of natural history, too many of them threw out the baby of humanity's classical intuitions of basic *spiritual* truths with the bathwater of outdated religious viewpoints.

'As a teen, I, myself, went through a time when I thought that if we couldn't see, or measure, something PHYSICALLY, it couldn't exist.

'I thought I was being rational, logical, *scientific*. I didn't realize, back then, that there are so *many* aspects of Life that cannot be reduced to physical measurements and MATERIALISTIC understandings alone. That there was, logically and necessarily, a role for *metaphysics,* for the *spiritual* and the *super* natural in our understandings.

'I guess you mean conditions and qualities like *Awareness and Mind, Love and Joy, Curiosity, and Creativity.* Like the *subtle Life Forces,* which the Chinese and Indian traditions refer to as *Chi or Prana, and, especially,* C O N S C I O U S N E S S itself.*'* Mused Alisha.

'Yes. Currently, mainstream, western science denies that *Consciousness* and *Mind* can be inherent principles of reality. That is, as ontologically real.

'Rather, it affects to see CONSCIOUSNESS and MIND as mere 'accidental byproducts' of a MINDLESS MATTER, which, it insists is the only 'stuff' that's truly real. According to materialism, if something, like the *Vital Forces, or Chi or Prana,* cannot be *physically* seen, or measured, it cannot exist. This is the narrow, inquiry-limiting, science-restricting, attitude.

'Yet, ironically, for this dismal, anti-spiritual worldview to *seem* true, we have to believe so much ***more*** **nonsense** than is ever asked of us by the world's religions.

'Because, despite their many faults, at least the religions don't insist that this world, and all in it, is just **pure, 'dumb' coincidence?**'

'Correct. This strange, anti-spiritual thinking, which currently so dominates our sciences, asks us to believe that *pure chance* is a supreme creative force—which is so obviously not rational—and it asks us to dismiss every single *spiritual, mystical, or psychical* experience, anyone has ever had, despite there being countless examples, just to try to make its own—not very well thought through—claims seem to be *'true.'*

'How many of us, Ali, ever stop to think that for 'MATERIALISM' to be *'true,'* we have to agree that every *Living thing, and* every *Living Being,* including, even, *our Children, our Pets, the lovely Flowers and Trees, the Birds and the Bees,* is, essentially, a mere chemical accident? *Seriously?*

'For so MATERIALISM to seem *'true,'* we have to convince ourselves that *Life* has no purpose, no intelligence, no meaning, no point.'

'It seems that for materialism to seem *'true,'* sighed Alisha, 'we have to make the worship of blind chance into our new (pseudo) religion.'

'Yes. Yet, ironically, materialism, as an ideology, is self-cancelling.'

'Why?'

'Because, IDEAS, and ideologies, Ali, like 'materialism' or 'naturalism' are not **physical** things, are they? Ironically, they emerge *from,* and they live *in,* the abstract realm of the MIND and the SPIRIT.

'They can be *represented* by symbols, by INK ON PAPER, turned from thoughts into poems, paintings, or equations represented by physical ink on PAPER, or into HOUSES, CARS, and PCs but, in essence, they are non-physical, mental, spiritual, ideational.

'Materialism is, **far from being scientific,** a dubious, inquiry-limiting, science-distorting ideology, a very poor set of bad *ideas.*

'So the sooner we move on from it, the better?'

'Yes.'

We continued on our way.

3 Darwin's New Law: Natural Selection

'For most of history, the idea that Life is just an accident which evolved by accident would have been thought laughable. But, after Darwin, it became mainstream—and the classical view became the minority view. Darwin sought to explain Life in entirely ***non*** *spiritual* terms—in the same way that geology was then being understood—as a purely physical process.

'Like others of his time, he had begun to doubt that all Life was *instantly* created, just a few thousand years ago. Geology had shown Earth to be far older than the mere six thousand years, proposed by Bishop Ussher in the 1600s, based on his far too literalistic readings of the timelines in the Bible.

'That was reasonable enough, wasn't it?'

'Yes. But Darwin didn't just switch to old-earth creationism, which allows for Life to be extremely ancient but still of *super* natural origin. No.

'What, he asked, if only the PHYSICAL was real? Ancient peoples would, like us, have wanted to make *sense* of things. However, not having our modern knowledge, maybe they invented various *gods* and *spirits* to try to explain their lives, and the natural worlds around them, investing the rivers and the seas, the mountains and the trees with *'Soul' and 'Spirit?'*

'Like imagining that the majestic thunder was the expression of this or that imaginary *'god' or 'spirit,' Life* a sacred gift, water holy and so on.

'Yes. *Or,* was it that like today's shamans, yogis, and sages, some of them were able to *directly* sense the presence of the deeper *Soul* and *Spirit* of things? They had *real* powers of *inner* seeing, of genuine *spiritual* vision?

'Most mainstream western thinkers of recent times have opted, like Darwin, for the first view. *'God' and 'spirit'* are wishful thinking. Maybe our modern prowess in explaining and harnessing the physical worlds to our uses, meant that we could, finally, abandon *spiritualism* as a way of understanding reality. If they were right, to continue to resort to *super* natural explanations *of Life and Mind* on our planet would no longer make sense.'

'Which, I guess, is why those who are **convinced** that materialism is true, that their worldview is *intellectually* justified, find it so hard to shift on all this. They are convinced they are *'right,'* so how can they shift?

'Yes. This is why, if you are a **materialist**, you *have,* like Darwin, to try to come up with purely physical theories about Life's history.

'Ironically, though, part of the enduring appeal of Darwin's theory has been that, if you are a *spiritual* believer, you can still, *kind of,* go along with it, because it is, clearly, more scientific than the creation story in the Bible.

'On the other hand, if you are *atheistically* inclined, you can allow the religious to keep their *delusions,* while **you** [think you] **know** that Darwin has shown us that *Life-Existence* is just a long-running cosmic accident...'

'Ok. So remind us, Ollie, of Darwin's basic idea...'

'Well, firstly, he assumed that Life had probably started as a random microbe of some kind. Then, he asked, *might* such a single cell organism, like a bacterium, *vary?, and vary?, and vary?,* until it slowly *'evolved,'* into multi-celled *Beings* like worms, or snails, or spiders?

'*Might,* at some point, a **two sex** world *'evolve?'* Male and female, with remarkably different parts? Penises for some, ovaries for others? [Dear Reader, how could this happen, by chance? Two entirely separate evolutions, based on random genetic mutations, **yet totally in sync**?! It doesn't make sense. 26]

'*Might* worms randoMly *'evolve'* into fish? *Might* fish slowly transform into amphibians? *Might* amphibians raNdomLy evolve into dinosaurs?

'*Might* shrews slowly *'evolve'* into sonar equipped bats? Might land-living bears gradually evolve into deep-sea diving, echo-locating, water-birthing whales, with highly specialized, collapsible lungs, enabling them to dive to amazing depths, 10,000 feet!, at truly incredible pressures?'

'Ok. But how did Darwin think **such fantastical processes** might happen, by pure chance? It seems to be such a shockingly unrealistic scheme . . .'

'Well, for example, Darwin wrote that he could, "see no difficulty in a race of bears being rendered, by natural selection, more and more aquatic in their structure and habits, with **larger and larger mouthS**, till a creature was produced as **monstrous as a whale**." [Dear reader, please glance at your book's cover as we ponder these, not one but, many quite extraordinary claims—backed up by no *extraordinary* evidence. Darwin's imaginings were not evidence.]

'But given the far-fetched, *'Just So'* story nature of some of these claims, I struggle to see how Darwin's ideas were any improvement on the classical hypothesis of *super* natural or *intelligent* causation?'

'Well, traditional religious creation stories were meant to *symbolize* the classical understanding that there is a *Living Intelligence, or intelligences* behind Living things, and behind *our* intelligences. They were never intended to tell us precisely how the process happened, in scientific terms.

'Darwin's theory, on the other hand, was attempting to fill both roles. (a) It implied that no *spiritual* or *super* natural intelligence was involved or *needed* to be involved in Life's creation, and its subsequent evolution. (b) It claimed to show, in a **literal**, and **scientific way** exactly *how* this, god-or-intelligence-not-needed, process had taken place.

'So it's as if Darwin was saying, "(i) *super* natural intelligence *is* not, involved in the *second by second* processes of Life's arising and its subsequent evolution, both now, and in the distant past. (ii) **I can show you** *why* it was not needed. It was because it unfolded in this very mechanical, linear way, which we can explain in terms of purely 'ordinary' or non *super* natural processes. Life is just another 'natural' phenomenon like any other."

'Yes, that's a good summary. But, without wishing to be rude to Darwin, it was delusional on his part. I can understand why many people, including Darwin, find Life difficult to understand, and Nature often very harsh.

'But, this does not prove that no intelligence was or is involved in the *moment by moment* arising of living things. It may, however, require a more sophisticated understanding of creation, than some wish to engage in.'

'How do you mean?'

'Well, for example, without opposites it would not be possible to have any creation at all. Without a down, how can there be an up? And without *sentiency,* that is without *subjective Beings* to know *subjective Being,* through the opposites, hot and cold, pain and pleasure etc, how can there be any *knowing Being* and the *knowing* of it? These are deep questions, which the mechanical, materialistic thinking of Darwin, completely ignores.

'Darwin was, however, very much speculating, around a **personal hunch**, which he was very excited about. It seems he truly felt that he might have hit upon a new natural law, one which explained very Life itself, and he hoped that, in due course, the fossil and other data would prove him right.

'His basic idea was that **all Life had descended, like a branching tree**, from just one or two, **universally shared, microbial ancestors,** due to their having **massively varied**, over time. He called his new law of random **variations to existing** organisms repeatedly arising, then, through breeding, multiplying, **Natural Selection**. An idea which sounds quite interesting, until we realize that, really, it is just a *tautology*, and all it is telling us is this:

> "That which survives, survives. Why does it survive? Because it is the 'fittest.' How do we know it's the 'fittest?' Because it survives. *Why* does it arise in the first place? Natural selection *cannot* tell us."

'Darwin referred to his scheme, as '**descent with modification.**' Because his idea was that all living things had descended from just one (or two) universally shared ancestor(s) cells, *versus* the traditional view that the species had been generated **separately** and were pretty much unchanging.

'By contrast, he asked, "What if the species could change more or less *indefinitely?* What if they were almost *infinitely* plastic, and **not fixed** at all?"

'What if *random* changes occurred in an *existing* organism? What if such changes were *beneficial* to the organism, *and* heritable? If so, its offspring should *do better* than the others, and *it's* descendants would predominate.

'What if a *microbe* could vary so much, due to *enough* random mutations [*millions?*], as to, eventually, transform into a worm, snail, spider, or fish?

'What if larger animals, like bears, *undergoing millions* of such random changes, very gradually, evolved into whales? [Please see your book's cover].

'Those organisms, which had so varied, with *heritable traits,* favoring their variant's survival, would leave *more offspring,* causing those traits to increase in the species—over the generations. So, for example, the more *whale-like bears* would, gradually, do better, in the seas, than the *less* whale-like bears. The *less spidery,* early spiders would not do as well as the later, more spidery, spiders. The earlier spiders silk would not work very well, but, those spiders, who had, due to millions of rAn*dom* changes, worked out how to spin *proper silk,* and to make *proper webs,* would, eventually, outcompete their less efficient counterparts. This was survival of the fittest.

'The *first* rodents would not be very *'bat-like,'* nor, would they be equipped with **sonar,** *far more efficient* than any sonar we make. But, gradually, over millions of random variations, with *innumerable* transitional versions, the highly efficient, sonar-equipped bats, we know, would triumph.

'This is why Darwin wrote that the **number of transitional species**, of bats, spiders, bears, whales, and so on—which ought to show this process at work—must have been *"inconceivably great."* [His words.]

'"Inconceivably great!" We need to imagine what that looks like. Vast numbers of 'trial' versions, *that failed.* Bears starting to become whale-like, some failing, some succeeding. Rodents starting to become bat-like, with many failures along the way. Only, **there's absolutely no evidence for them**!'

'In ***Darwin's Enigma,*** Luther Sunderland explores these issues. He writes "Geology and paleontology [FOSSILS] held great expectations for Darwin, although in 1859 **he admitted** that they [also] presented the STRONGEST SINGLE EVIDENCE AGAINST HIS THEORY. [The] **Fossils** were a perplexing puzzlement to him because they **did not reveal any evidence** of a gradual and continuous evolution of life from a common ancestor, [the very] **proof** which he needed to support his theory."[27] Ah, Alisha, we need to let that sink in!

'So how did Darwin deal with this? It disproved his theory, didn't it?'

'Well, he just hoped that the vast numbers of transitional or linking fossils needed to prove his theory right would, eventually, be found.'

'Ok, fair enough. So, were they ever found? That seems pretty key . . .'

'No, they were not. But before we go further into that, let me finish my summary of Darwin's basic idea. Namely, it was that once his *hypothetical,* universal ancestor microbe had begun to vary and to *'evolve,'* it would, perhaps, do so in different directions. For example, might one bacterium, slowly, become a worm? First offshoot. Another an ammonite? Another offshoot. Another a fish? Each **new anatomical design** would be a new offshoot. These offshoots would be the first branches on his imagined tree of evolution. The trouble is, as we'll see, **Darwin's tree has *never* been confirmed by the real world data of the fossils** and molecular biology.

'But some of these ideas, like the bears, seem utterly **laughable**, Ollie. What about Darwin's scientific colleagues? Did they take them seriously?'

'No, most of them did not—far from it. Despite not having our modern knowledge of the truly mind-boggling complexities of living things, there were plenty of scientists in his day, who were skeptical of his claims, which, they thought, were **unrealistic**, and **unsupported by evidence**, with many obvious flaws. For example, St. G. Mivart, who published *On the Genesis of Species*, wherein he *doubted* that natural selection could **initiate** biological structures. According to S. J. Gould, a highly regarded modern fossil expert,

> 'Darwin offered strong, if grudging, praise and took Mivart far more seriously than any other critic... Mivart gathered, and illustrated "with admirable art and force" (Darwin's words), all objections to the theory of natural selection. . . "a formidable array" (Darwin's words again). Yet one particular theme, urged with special attention by Mivart, stood out as the centerpiece of his criticism. ... **No other criticism seems so troubling, so obviously and evidently "right"** [yes!] (against a Darwinian claim that seems intuitively paradoxical and improbable). Namely, as Mivart referred to it: "**The Incompetency of 'Natural Selection'** to account for the Incipient Stages of Useful Structures." [28]

'Mivart realized that ***natural selection*** cannot *initiate* living things?

'Yes, Because it has *no reason* to select any chemicals in the first place, let alone the **very specific** combinations needed. Darwin based his ideas partly on his knowledge of the changes that *intelligent* people, with *intelligent aims,* could bring to an *existing* species. Humans who *already* had something *very cleverly* put together, like wolves or pigeons or tulips, to select from. Whereas, evolution by natural selection relies on **rAnDo*m*** changes for all its creativity.'

'But is there any *reason* for such an evolution ***even to begin?***'

'***No,*** of course not. ***How*** *could it?* ***How*** *could* a **DNA CODE**, or ***Cells,*** arise by chance? Natural selection, nature 'choosing' the 'fittest' variations to existing species, ***cannot explain the very start of Life.*** Because, at that point, there are no *'fit' or unfit* species to vary, or, to choose from, are there?

'There are no *Living* things ***to*** survive. Nature's selection, via the superior survival rates of the *'fittest'* variations, can only *refine,* to a limited degree, what *already* exists. It cannot generate the truly new. How can it? The creation of any complex mechanism, including *Living things,* needs *mind and imagination.* But Darwin's *'evolution' has **no** mind.* This is what MATERIALISM means: *all of Life* is just a series of chemical flukes.

'It doesn't add up. But materialists say they can't believe in miracles. Yet, think about it, *every* living thing is a *miracle: Cells!, blood!, DNA Codes!, Hearts!, Lungs, Eyes, Minds!, Spiders, Bats, People.* Pure chance? Honestly?

'So, the idea that Living things *did* arise, and *did* become ever more sophisticated, by pure chance, has *never* had a justifiable *scientific* basis?'

'No, absolutely not. It is simply the view of those who, due to their dislike of the religions, and due to the *physical invisibility* of Life's deeper causes, are unwilling to use classical *logic,* and commonsense, to *infer* that their own, *bright* existences, are more likely to have *bright* causes than MINDLESS. For, they say, "If we cannot *physically* see it, it cannot exist!"

'Yet, most of those, who think this way, seem blissfully unaware of *just how much nonsense,* **how much utter nonsense** they, and we, need to believe, in order to make their belief system *seem* true. But, if we point out that their scheme is *unscientific,* horribly so, that it is deeply irrational and illogical, that it is, in fact, demonstrably *impossible,* they ignore us.

'Unfortunately, once Darwin's ideas caught on, too many thinkers did not notice that their own—newly adopted—beliefs in *blind chance,* as a supreme creative force, were, **oh how ironic!,** a far weaker, and a far more *illogical* way of thinking than some of the old religious and traditional spiritual ideas which they wished to replace.

'At least the religions understood that it makes much better sense, overall, to hypothesize that *intelligent Life* emerges, from a matrix, which is, itself, *intelligent.* The idea that mere luck can create anything remotely clever has **never** had a *scientific* foundation. It was just a new *faith.*

'Some scientists did question this new *faith.* They argued that natural selection could only account for the survival, or minor variation, of what *already* so cleverly existed. It could not account for its arrival in the first place.

'In **Darwinism: The Refutation of a Myth**, Soren Lovtrup wrote, "some critics turned against Darwin's teachings for religious reasons, but they were a minority; **most** of his opponents ... argued on a completely scientific basis."

'Lovtrup added that "the reasons for rejecting Darwin's proposal were many, but, first of all, that many innovations *cannot possibly* come into existence through the accumulation of many small steps... because incipient and intermediate stages are **not [in any way] advantageous**.'[29]

'So, what Darwin called 'natural selection,' cannot *initiate* Life?'

'No. Natural selection can only favor *existing* living things. For example, finches born with *larger beaks*, doing better, in certain climatic conditions, than finches with smaller beaks. But natural selection has neither the ability, nor any reason [as it is mindless], to 'select' the very ***complex*** mixes of chemicals needed for living things—like finches—to arise in the first place.

'As a MATERIALIST, Darwin *had* to see Life as 'JUST CHEMICALS,' nothing else. Materialism *has* to deny the classical hypothesis that there may be ***subtle, non-physical,*** components to the creation—and to the *second!, by second!,* maintenance—of all *Living things,* like the *Vital or Soul Forces, animating all Living Beings*—versus MINDLESS DEAD MATTER.

'Yet, as Michael Flannery writes,

> "By and large, the scientists of his day were not much impressed with Darwin's theory. John Herschel called natural selection "the law of higgledy-piggledy," and William Whewell thought the theory consisted of "**speculations**" that were "QUITE UNPROVED BY FACTS," so much so that he refused to put the book on the shelves of the Trinity College Library."
>
> "Rather, it was the reading elite of London that was captivated with Darwin's theory. "Freethinking" bohemians and assorted **society trendsetters** grabbed up copies of the *Origin* and later *Descent*. The secular [atheistic] creation myth they had long been looking for was finally in hand. ... In the end, it was Darwin's rhetorical **salesmanship** that won the day ... his sheer presentation. It was a pitch easily made because it "sold" a product the intellectual elites had long been waiting for, **a theory of life in which God [the *super* natural]** was superfluous and irrelevant." [30]

'Darwin's —"Nature selects the best random variations"— idea, which he was so taken with, *cannot build the functional* from nothing. It is just a scientific sounding *"Just so"* story. Because, all natural selection can do, is favor those—*existing*—species members best suited to the current conditions.

'For example, the offspring of the bears who are, genetically, the best swimmers, will *not* become whales, but they will, generally, survive better than the offspring of the bears who are more genetically feeble.'

'Darwin knew, perfectly well, that *no species* had ever been bred, even by the most skillful breeders, using very careful artificial selection, to vary beyond a certain point without damage. Yet, ignoring the INABILITY of NATURAL SELECTION to ORIGINATE ANYTHING FUNCTIONAL in the first place —the *actual* origin of species—Darwin argued that a random change in an *existing* plant or creature might arise, it might survive, and, via an "**inconceivably great**" number of transitional versions it might lead, eventually, to a new species, then another, and another. Only, the fossil, and other evidence, **has never supported Darwin's speculations**.

'It seems there was a *lot* of SPECULATION on Darwin's part?'

'Yes. Darwin's ideas *were* always highly speculative. As Whewell said, Darwin's hunch, his beloved theory, was "QUITE UNPROVED BY FACTS."

'In ***Darwin's Bluff, the Mystery of the Book Darwin Never Finished***, Robert Shedinger reminds us of Darwin's own, self-exculpatory words, from the beginning of *"On the Origin of Species."* Darwin wrote, in 1859,

"My work is now **nearly finished**; but *as* it will take me two or three more years TO COMPLETE IT, and *as* my health is far from strong, I have been urged to publish… SOME BRIEF EXTRACTS from my manuscripts.

"This ABSTRACT, which I now publish, must necessarily be imperfect." For, he wrote, "**I cannot give references** and authorities for my several statements…" This was useful for Darwin—if he was *bluffing* as to the real existence of **supporting evidence** from the fossils. So, playing for time, he wrote, "I can here give only the general conclusions at which I have arrived, with a FEW FACTS in illustration, but which, I hope, in most cases will suffice. No one can feel more sensible than I of the **necessity** of hereafter PUBLISHING IN DETAIL ALL **all the** FACTS, WITH REFERENCES, on which my conclusions have been grounded; and I hope in a future work to do this. For I am well aware that **scarcely a single point** is discussed in this volume on which FACTS cannot be adduced, often <u>apparently</u> leading to **conclusions directly opposite** to those at which I have arrived.

"A FAIR RESULT can be obtained only by fully stating and balancing the facts and arguments *on both sides* of each question; and this cannot possibly be here done." [31] Again, convenient for Darwin. Why did he not "state the arguments on both sides?" We also must question his use of the word '<u>apparently</u>,' because, if we do look at "both sides," a "fair result" argues, far more strongly, *not* for the descent of all living things from a universally shared microbe ancestor, like a bacterium, but for their *separate* descents.

'That's intriguing, Ollie. So, did Darwin, later, publish, "**in detail**," "ALL THE FACTS, with references" on which his "conclusions were grounded?"

'No. He didn't. That's the point. Yet, in his subsequent exchanges with other scientists, Shedinger notes how Darwin **repeatedly** told them that *"On the Origin"* was **but an introduction** to his ideas, which he *would* later substantiate in a much more weighty and more authoritative work.

'Strange, because he had said, "My work is now **nearly finished,**" and that it would take him just "**2 or 3 more years**" to complete it." So what happened?

'Two or 3 years is not long, and he said his work was 'nearly finished.' So his more authoritative work, backed up by far more **facts**, illustrations and REFERENCES, should have been out by around 1862. It never came. Yet he lived on, for *another 20 years*, until 1882. His work was 'nearly finished' but all we the public, and his fellow scientists, ever got to see were those few "brief extracts" from his allegedly much larger work. A *bluff?*

'Of course, this would not matter *if* the ideas he had outlined in *'On the Origin'* had, in time, been borne out by the *actual evidence* of the fossils, as Darwin hoped might happen. But, despite all the claims, they never have.

'Ignoring the lack of the real world data, he needed to support his theory —like fossils of **bears slowly becoming whales**, shrews slowly becoming bats, un-spiders gradually becoming spiders—Darwin just hoped that he might have hit upon one of reality's *key* organizing principles, a new natural 'law,' almost like gravity, and that only the *details* remained to be filled in.'

'So, do you think that Darwin was *bluffing,* Ollie? It's a serious charge.'

'Perhaps not deliberately. More the bluff of someone who is very *excited* about their scientific idea, who *believes* they are right, despite the lack of confirming data, but who hopes that, *in time,* they will be proved right.

'Except, **the supporting data, Darwin needed, have never been found?**'

'Correct, and this is where we, the general public, and, indeed, many scientists, who are not specialists in the field, have been misled.'

'So, the fact that Darwin's idea was interesting, did not make it **true**?'

'Of course not. 'Interesting' does not, of itself, make something true. Yet, we have endlessly been assured that Darwin successfully explained Life's history. We have **not** been told, the far more accurate, and far more **truthful**,

"It was a scientific hypothesis of its time, but it did ***not,*** in the end, prove to be correct." Darwin's supporters have never accepted this.

'They have, **consistently, refused** to allow the *real-world* data to speak, and to let Darwin go. Much as Darwin tended to do, they seem to feel that his ideas are *'too good'* not to be true. So, *'they must be right!"*

'But that's not science, Ollie, it's a *faith* system… a kind of religion.'

'I agree. Bear in mind, though, that there is a lot of *politics* and *philosophy* involved in this area. It is naive to think that all this is just about pure science, and the dispassionate pursuit of truth. Because many people now see science as 'RATIONAL' and spirituality as *'irrational,'* they would rather pretend that Darwin's theory is correct than risk letting *Soul and Spirit* back into our understandings of Nature—if they admit his theory has failed.

'It *is* true that Darwin's theory was elegant and it was clear: original, universal, ancestor-microbe, to vast, spreading, Tree of Life. What a simple, clear, logical, MECHANICAL, idea. One which *religious* believers could argue was *God or Spirit's* way of creating, and his materialist and *atheistic* fans could argue was a purely rANdo*m* process.

'But the simplicity of Darwin's basic idea of evolution, microbe to worm to fish to bear to whale—rANdo*m* or **intelligently** driven—did ***not*** make it true. The data never obliged *either* his materialist fans, or his more *spiritually* inclined followers. This explains why he never published his much promised, later, larger work. Darwin lived on for 20 years, past his just '2 or 3 more years' deadline to complete it. Yet, like his supporters, then and ever since, Darwin was reluctant to admit to the failure of his theory. This is why his fans, like the unhappy sisters in *Cinderella,* continue, to this day, to try to force the *'slippers'* of the real world facts of the fossils, embryology, genetics and molecular biology onto the hapless *'feet'* of his initially intriguing, but, now, thoroughly disconfirmed, ideas.

'So the superior survival rate of the 'fittest' can only *refine* what already, so cleverly, exists? *It cannot explain* the arising of Cells, or Blood, or DNA Codes, or Livers, or Lungs, or Bones, or Brains, or Microbes or Worms or Spiders or Fish or Bears or Whales or People in the first place?' Queried Ali.

No, how can it?
The MINDLESS Nature of
materialism cannot create anything…
Natural selection cannot account for Cells.
Natural selection cannot tell us how Life begins.
Natural selection cannot, explain Life's DNA CODES.

Natural selection *cannot* create CODE. It has no reason to. CODES require *imagination* and intelligence to be created into reality. Natural selection *cannot* account for *intelligence and purpose* in nature. Natural selection *cannot* explain the geologically *sudden appearances* of the many weird and wonderful animals of the Cambrian Explosion. Natural selection *cannot* explain the existence of *Mind or Thought.* Natural selection *cannot* explain *goals* in minds. *It has no mind.* Natural selection *cannot* explain *sentiency, senses, or feelings.* Natural selection *cannot* cause an existing species to transform into one wholly new; and there is *no evidence* that it has ever done so.

Yet, natural selection was
Darwin's core idea.

'So Darwin's hypothesis was never borne out by the real world data…'

'Correct. The devil *was* in the DETAIL, which, as we will *repeatedly* see, neither Darwin, nor anyone since Darwin, has ever been able to provide.'

'In *Natural Selection or Natural Creation? – A Complete New Theory of Evolution,* John Davidson discusses the **Hindu / Yogic** concept of creation as an Eons Long Creation in Intelligent Evolution (ELCIE). Here, he notes just *some* of the many fatal flaws in Darwin's atheistic scheme. He asks,

"What happens, then, if one attempts to understand this vision of nature's wholeness in terms of conventional [Darwinian]

evolution theory...? One is immediately led to wonder **how such wholeness** could have arisen from linear and **random processes?** ...

"Let us take a specific example: the evolution of a spider. The question is: **how could** a spider have evolved by **random mutation**?

"A spider exhibits a host of integrated biochemical, ... anatomical, behavioral and instinctive adaptations. It is able to manufacture **perfect silk**; to spin and weave it into perfectly organized webs and traps; to know how to walk in a sticky web without getting entangled; **to know** how to wrap up a fly, but to identify and **leave a wasp alone**; to ... [wait] patiently at the edge or center of the web; **to paralyze prey with so perfect a toxin** that it does not then poison itself when it sucks out the body juices with specialized mouth-parts – and so much else besides . . .

"So, musing upon how a spider could have evolved so many integrated skills and biological systems from some unspiderly ancestor, **are we to suppose** that the earliest spiders **wove rather poorly structured webs** out of inferior silk; got stuck occasionally in their own webs; could not distinguish between dangerous hornets and houseflies; **used poor toxins which only partially killed or paralyzed their prey** while later giving them a tummy ache or worse; and that they were ill adapted for actually eating and digesting their prey and so they only relied partially upon their spiderly ways for making a living, anyway? **In short, that they were pretty much of a mess,** still needing considerable time, not to mention **several million "chance" mutations** [which are rare events and all cells actively protect against], to get their act together?"

'What Davidson says makes sense to me, Ollie. It seems obvious.'
'I agree. He goes on to make further powerful points. He writes,

"Since such evolution is still supposed to be in progress, why are there **no semi-evolved creatures** living NOW? **Why are there none found in the fossil record?** In fact, 300 million-year-old spiders with web-weaving spinnerets have been found in the fossil record, YET NOWHERE has any **fossilized species been found** which is ON THE WAY **to becoming a spider.** It's spiders or something else! All creatures, not just spiders, are well structured wholes... Evolution by chance and by miniature steps would pull the creature in a host of different directions... For one **cannot attribute any goals [or progressive quality]** to a materialistic view of evolution.

"Nature cannot be assumed to KNOW WHERE SHE'S GOING in such a philosophy of chance and chaos. And in the absence of any consciousness, vision or volition, how could such perfect designs as we see around us have come into existence?

"Yet somehow we are ASKED TO BELIEVE that living creatures, who most clearly possess volition, desires, goals, *MIND* and *CONSCIOUSNESS* have arisen by random processes from DUST and WATER which exhibit none of these.

"Materialistic evolutionary theory is forced to assume that there is **no apparent purpose** to life and that natural selection is *not* selecting for some future, *intelligently* conceived goal; it is a selection only conferring advantages in the current moment.

"**How then** could some radically new species design [microbe, worm, snail, spider, fish, bird, feline, canine, ape, human] **emerge purely by chance,** and over an immense span of time, yet exhibit such a multitude of **integrated characteristics?**" [32]

'In 1998, molecular biologist Michael Denton wrote *Nature's Destiny: How the Laws of Biology Reveal Purpose in the Universe.* He wrote,

'The **design of living systems**, from an organismic level right down to the level of an individual protein, is so *integrated* that most attempts to engineer even a relatively minor functional change are bound to necessitate a host of subtle compensatory changes. **It is hard to envisage a reality *less* amenable** to Darwinian change via a succession of independent UNDIRECTED mutations altering one component of the organism at a time.' [33] [Another who is thinking.]

'This is crucial. *Millions,* not just a few, of integrated biochemical changes would be needed for darwinian-style evolution to take place.

4 Where's the Tree?

'Darwin wondered if it might be possible to explain all of Life's history in *non-supernatural* terms, starting with a ra*ndo*M microbe which, having arisen by chance, *might* vary so much, as to, slowly, turn into a vast Tree of Life.

'Might the first microbial Life, like a bacterium or alga, vary so much as to provide an *offshoot,* then more offshoots, until, eventually, at the branches tips, would be found all today's species? It was an interesting idea. But the massive problem with it is that the **fossils have never agreed** with Darwin's key idea of one species gradually changing into another by, as he wrote,

> "**differences not greater than** we see between the **natural** [wolves, wild sheep, tulips] and **domestic varieties** [dogs, sheep, tulips] of the same species at the present day. ... [where] the **number of intermediate and transitional links**, between all living and extinct species, must have been INCONCEIVABLY GREAT. But assuredly, IF THIS THEORY BE TRUE, such [inconceivably great numbers of transitional species] have lived upon the earth."34

'Ah yes… if this theory be true. As a teenager, I took what I learnt about Darwin and his theory **on trust**. But, later, when I looked a little closer, I was surprised to discover that the **all-important fossils** did not support his ideas.

'Not only are there *no "inconceivably great"* numbers "of intermediate fossil links, between all living and extinct species," his predicted transitional species are never to be found. Yet, *millions* of us *still* believe in Darwin's theory. It is one of the great curiosities and oddities of our time.'

'In ***Darwin's House of Cards*** Tom Bethell recalls how he once toured the Natural History Museum in London with its then senior fossil expert, Colin Patterson. Patterson told Bethell "that he was looking for cases where the *(hypothetical)* shared ancestor of two later species was identified in the diagrams on display." If they had ever existed, they would be found at the nodes or branching points, on Darwin's *hypothetical tree* of life. But, as

Patterson pointed out to Bethell, Darwin's *hypothetical* branching points on his hypothetical tree *were all **non-hypothetically empty**!* Darwin's imagined shared ancestors were nowhere depicted, because they are never to be found.

'Patterson told Bethell that, "as far as he could see," the nodes, the places where Darwin's common ancestors *ought,* in Darwin's theory, to be, "are always empty in the tree of life," and he doubted that they'd ever be filled.

'So Darwin's notion of random, single-cell-microbe, to vast, tree-of-Life, remains, even now, to this day, just a hypothesis, not one ever proven true?'

'It is worse than that, Ali. It has been FALSIFIED, repeatedly. But this is not yet much conceded, at least not to the general public. However, rather than admit to this failure, some of Darwin's modern supporters say that the evolution they *now* believe in is ***not*** tree-like—but network-like.'

'But that's not Darwinian, is it?'

'No, it isn't. It contradicts Darwin's *key idea* of common descent, with the species, he argued, *able to vary*—more or less *without limit*—during this very long process: Bacterium to worm—*huge* change. Worm to Fish—*massive* change. Dinosaur to Bird—*huge* change. Bear to Whale—*massive* change. While Darwin was entitled to his hunch or bluff, we need to think all this through. **Is his scheme plausible**? Is there any good evidence for it?'

'So why did Darwin argue for such an *extreme* process?'

'Mostly, because he wanted an alternative to the classical hypothesis of *intelligent* creation. If you think of *spiritual* intelligence as an Elderly Man, with a long, flowing beard, as, historically, some of the less educated had done, then, understandably, you may want to look for something different.'

'But that's not how I think of *God* or *Spirit* Ollie! I think of *It/S/he* or *ISH* as this amazing, *'Living, Loving, Intelligence'* which is *in* everything, which *in-forms* everything. I don't think of it as a bearded god figure!'

'I agree. I think holistic reality is the ever-fascinating *Expression* of an *amazing, Self-existing, Intelligence which with infinite mystery, **just Is**. And, right now!, this second!, and every second!, this same, mysterious intelligence* is making my heart beat, and yours, my blood circulate, and yours. From the micro, in.form.ing every cell in our bodies, to the macro, turning the planets . . . circling the stars . . . swirling the galaxies. Not just *generator* of Nature's Laws, *It/S/he or 'ISH'* ***is*** the Laws. *The Stupendous, Absolute Existence,* from which *All things* emerge, and to which, eventually, all things, from atoms to galaxies, return—call *Ish God,* call *Ish Allah,* call *Ish Brahman,* call *Ish Buddha Nature or the Tao,* call *Ish the Great Spirit or, simply, the Great Mystery,* the name doesn't really matter . . .

'But Darwin was not keen on that idea?'

'Well, he was of his time, and many, in the west, still had rather naive ideas of *Spirit.* So, for the more scientifically minded, like Darwin—but not aware, for example, of the *Hindu and Yogic* traditions, which have a much more sophisticated understanding of cosmology than the western tradition—those ideas no longer made much sense. But, rather than considering a more *nuanced* view of *Spirit,* Darwin *confessed,* in a notebook, that he was "a **materialist**," although, he added, he did not "dare say so openly."[35]

'So, he tried to find a purely PHYSICAL explanation for Life's history, in the form of his **idea of descent** from a universally shared ancestor microbe plus his branching tree idea. We, though, now know so much more than Darwin and his early followers did. **We cannot scientifically justify** their idea that all Life on our planet started by mere chance, in some lightning-hit sea.

'This idea, while still popular, today, among many who see themselves as the thinking classes, is no longer intellectually credible, no matter what famous contemporary Darwinians, like Richard Dawkins, may have to say.

'In his book, *Darwin's House of Cards,* Tom Bethel refers to "DARWIN'S MISTAKE" He writes:

> 'We all know that small differences are observed between the generations. [For example finch beak variations and variably hued moths of the same species.] ... Darwin's mistake was to assume that these [minor] differences **somehow accumulate** over the millennia, **so that one species eventually transforms itself into another**.
>
> 'WITHOUT EVIDENCE Darwin's supporters today *still* [believe this]. ... But such a transformation HAS NEVER BEEN OBSERVED [either in living history or in the fossil records]. **No species** [fossil or living] **has ever been seen [or known] to evolve into another**.'[36]

'Some people refer to these minor differences between the generations as micro-evolution. But it is a *misleading* term. Because these changes tend to be *directionless* and *reversible*. For example, the proportions of finches with different sized beaks, larger, smaller, varies from year to year, as conditions change. But these variations are *not* cumulative. They provide no support, at all, for the type of large-scale evolution which Darwin proposed.

'In an article, titled 'Demolishing Darwin's Tree,' *Evolution News (EN)* comments on a paper by researchers who are, "ready to provide a more "expansive" view of evolution that replaces Darwin's tree with a *"network"* of life. Why? Because "genetic data are not always tree-like."'

'But, *EN* asks, "If they replace the tree with a complex set of inter-connections, what happens to" Darwin's iconic "notion of universal common descent?" *Evolution News* quotes from the paper,

> "However, the more we learn about genomes **the *less* tree-like** we find their evolutionary history to be . . ."

'*Evolution News* remarks: 'Interesting: if "evolutionary history" is *not* **tree-like,** does [Darwin's key idea of] **universal common ancestry** [of descent with *vast* changes, into a branching tree of life] still hold? ... Calling it "historic," the authors recognize the extent of the shift they are proposing:

> "... it is a shift away from strict tree-thinking to a more expansive view of what is possible in the development of genes, genomes, and organisms through time."

'*Evolution News* comments on these major shifts in Darwinian thinking: "They use [the words] "development ... through time" as a SYNONYM for evolution. But what kind of evolution? If it is *not* **tree-like,** what is it? In a network diagram, common descent gets scrambled... with no clear progression from simple to complex. ... There's nothing here about a beginning and a progression...[37]"

'But a 'beginning and a PROGRESSION' is key to Darwin's idea of shared descent, isn't it, with *huge* changes, all along the way, in organisms like Bacteria, leading, eventually, he said, to organisms like Bears or Whales?'

'Yes. If we drop Darwin's tree, his *'hunch'* is over. His tree *is* key.

'Yet, as biochemist Michael Behe points out, "Darwin's "tree of life" – where a *single, primordial cell* gave rise to all subsequent organisms – *is dead*. The DNA sequence data *cannot* be made to fit with the idea." [38]

'What most of us don't realize, Ali, because we're not told, is that the fossils **never** supported Darwin. Rather, when new species arose, they did so **abruptly**, in a mysterious pattern of **sudden** emergences.

"In any local area, a species does *not* arise gradually, by the steady transformation of its ancestors; it appears **all at once** and **fully formed**."[39]

'So wrote, world famous fossil expert, S. J. Gould. Earlier than Gould, another world class fossil expert, George Gaylord Simpson, stated,

> 'This is true of all thirty-two orders of mammals . . . In most cases the BREAK IS SO SHARP and the G A P so large [between Darwin's alleged common ancestors and their descendants] that the ORIGIN of the order is SPECULATIVE and MUCH DISPUTED...'
>
> 'This regular **absence of transitional forms** is not confined to mammals, but ... is true of almost all classes of animals, both vertebrate and invertebrate ... [It is true] of the major animal phyla [basic animal designs, like *Mollusca*, snails, *Nematoda*, worms, *Arthropoda*, spiders, *Insecta*, ants, bees, beetles, butterflies. *Crustacea*, shrimps, *Chordata*, fish, amphibians, reptiles, birds, mammals, humans], and it is apparently also true of analogous categories of plants.' [40]

'Simpson, strangely, then ignored the Darwin-negating implications of his own research. He wrote: "Man is the result of a *purposeless* and natural process that did not have him in mind." How did he *know?* He didn't know.'

'So, despite all the claims, there is **no good fossil data** to support Darwin's idea of a universally shared descent—microbe-to-fish-to-man?'

'No, there is none. This is one of the *very odd* things about this debate. When, after my own time as a Darwin fan, I began to doubt that *any* entirely *non-spiritual* theory of life, including his, could fully explain the complexities of Life on earth, I did not realize just how much the actual facts of natural history, including the fossils, did not bear out his ideas at all.'

'But it is so odd, Ollie, in view of all the pro-Darwin claims . . .'

'I agree. It is also hard to believe, in light of all the pro-Darwin propaganda. But, sorry to say, it is true. In ***Darwin's Enigma***, Luther Sunderland explores these troubling issues.

"Geology and paleontology [FOSSILS] held great expectations for Darwin, although in 1859 he admitted that they [also] presented the STRONGEST SINGLE EVIDENCE AGAINST HIS THEORY. [The] **Fossils** were a perplexing puzzlement to him because they **did not reveal any evidence** of a gradual and continuous evolution of life from a common ancestor, [the very] **proof** which he needed to support his theory.

"Although [the] fossils were an enigma to Darwin, he ignored the problem [of the **lack of supporting data**] and **found comfort in the *faith*** that future explorations would reverse the situation and ultimately prove his theory correct. He stated in his book, *The Origin of Species*:

> "The geological record is extremely **imperfect** and this fact will to a large extent explain why **we do not find** intermediate varieties, connecting together all the extinct and existing forms of life by the **finest graduated steps**" ["Finest graduated steps?! *Seriously?!* I'm sorry, but this was either a bluff or delusional of Darwin. *Why?* Because, we do not find *any* such fossils at all!" He added:] "He who rejects these views on the nature of the geological record, WILL RIGHTLY REJECT MY WHOLE THEORY." [41] [Here, at least, he was correct.]

"Now," Sunderland continues, "after **over 120 years** of the most extensive and painstaking geological exploration of every continent and ocean bottom, the picture is infinitely more vivid and complete than it was in 1859. Formations have been discovered containing hundreds of **billions of fossils** and our museums now are filled with over 100 million fossils of 250,000 different species. The availability of this profusion of hard scientific data should permit objective investigators to determine if Darwin was on the right track.

"What is the picture that the fossils have given us? **Do they reveal** a continuous progression **connecting** all organisms to a **common ancestor**? With every geological formation explored and every fossil classified it has become apparent that these, the only direct scientific evidences relating to the history of life, **still do not** provide any [of the fossil] evidence for which Darwin so fervently longed.

'The G A P S between major groups of organisms have been growing *even wider* and more undeniable. They can no longer be *ignored* or rationalized away with appeals to the imperfection of the fossil record. [As Darwin tried, and his followers, *still,* try to do]...

"By the 1970s prominent scientists in the world's greatest fossil museums were coming to grips with the so-called "gaps" in the fossil record and were cautiously beginning to present *new theories* of evolution that might explain the *severe conflicts* between neo-Darwinian theory and the hard facts of paleontology [fossils].

"Back in 1940, Dr. Richard B. Goldschmidt had faced the horns of this dilemma-of-the-gaps with his *hopeful monster* theory, the idea that every once in a while an offspring was produced that was a [GAP jumping] monster grossly different from its parents.[42]

"Goldschmidt's . . . ideas were ridiculed for many years . . . [However, in the 1970s] Niles Eldredge, curator of Invertebrate Paleontology at the American Museum, was collaborating with Dr. Stephen Jay Gould of Harvard, and calling their **new theory,** aimed at explaining the G A P S "punctuated equilibria." [PE]'[43]

'Eldredge and Gould, like Goldschmidt, were trying to make sense of the fossils as they really are, without, however, being prepared to admit that Darwin's theory had failed. They were trying to explain the **sudden**, non-evolutionary, **non-Darwinian,** appearances of new species, *without any,* let alone the vast numbers of transitional fossils which Darwin predicted.

'So, Darwin's famous theory of evolution, which we are all supposed to believe in, and be thought ignorant if we do not—Bacterium to Bear, Bear to Whale, Rats to Bats, Particles to People—has never been true? It is a huge HOAX? A joke? Just a scientific sounding *"Just So"* story?'

'Not exactly. Because those who believe in it are, I hope, sincere. Ironically, I think many of us HOAXED *ourselves* into believing Darwin's idea of a shared, tree-like descent, Bacterium-to-Worm-to-Fish-to-Bear, Bear-to-Whale, Particles-to-People. We saw what we *wanted* to see, not what is there. SUDDEN appearance is the rule, *not* Darwin's predictions. This is why trying to force the data of the UNHELPFUL fossils, and the UNHELPFUL DATA from embryology, genetics, and molecular biology, to fit Darwin's theory, is such a waste of time. I just hope that, gradually, more of us will begin to **see** what, before, we did not see. That Darwin's theory has never been borne out by the data, and it is now, just a *faith*.

'Like the *Emperor's New Clothes,* we need to call it out. Because, there is **no evidence** (at all) that single-cell microbes gradually turned into multi-celled worms, or spiders, or fish. Because, there is **no evidence** that Fish gradually transformed into Frogs, Fish-frish-frosh-frogh-Frogs.

Because, there is no evidence that Dinosaurs gradually turned into birds. Because, there is no data that land mammals, like Bears, slowly turned into sea going, echo-locating Whales. There is no evidence that land-running rodents, like Shrews, gradually turned into *fast-flying,* sonar-guided Bats.

'No, what the fossils *actually* show us is that, in radical contrast with Darwin's ideas, the new species appeared **abruptly**, in a very non-Darwinian and in a very **non-evolutionary** way. Then, in further conflict with his claims, they *stably* lasted, *without* changing into anything else, before, eventually, disappearing into extinction.

'Darwin was *aware* that the fossils did not support his theory. He just hoped that, *eventually,* he'd be proved right. As Shedinger says in ***Darwin's Bluff,*** Darwin was so attached to his theory that he was loath to let it go.

'I guess he was very passionate about his ideas, and their potential to gift him lasting scientific glory, if he really had unravelled the mysteries of Life on earth? So, he just hoped that the "inconceivably great" numbers of fossil linking species his idea called for, would, eventually, be found?'

'Yes. Shedinger thinks that Darwin's ego got in the way of the data. But, admitting at least *some* of the problems with his ideas, Darwin wrote,

> "LONG before having arrived at this part of my work, a crowd of difficulties will have occurred to the reader [**Yes!**]. Some of them are so grave that to this day I can never reflect on them without being staggered [**Yes!**]; but, to the best of my judgment, the greater number are only apparent [**No.**], and those that are real are not, I

think, fatal to my theory.' [This was **delusional.** So much is fatal to Darwin's theory. The refusal to accept this is *ideology,* not science.]

"These difficulties and objections may be classed under the following heads :- Firstly, *why,* if species have descended from other species by insensibly fine gradations, do we not **everywhere** see innumerable transitional forms? ["**everywhere**"? **Nowhere!**] Why is not all nature in confusion instead of the species being, as we see them, well defined [and in separate kinds]? [**Correct**.]

"Secondly, is it possible that an animal having, for instance, the **structure and habits of a bat,** could have been formed by the modification of some animal [like a non-flying rodent!] with wholly different habits? [The answer is "no." There's **no evidence for any such process.**] **Can we believe** that natural selection could produce, on the one hand, organs of trifling importance, such as the tail of a giraffe, which serves as a fly-flapper, [**not trifling to the giraffe**] and, on the other hand, organs of such wonderful structure, as the eye, [**No, we cannot, realistically, believe that**.] of which we hardly as yet fully understand the inimitable perfection? [*Yes.*]

"Thirdly, can instincts be acquired and modified through natural selection? What shall we say to so marvelous an instinct as that which leads the bee to make cells, which have practically anticipated the discoveries of profound mathematicians?'[44] [Well, the materialist school does assert that Beehive making arose by chance. Spiders webs arose by chance. Eggs arose by chance...]

'So Darwin had his own doubts?' said Alisha, looking surprised.

'Yes. If we deny any part to the *super* natural, in the creation and maintenance of nature, *this second!, and every second!,* we *have* to attribute truly irrational abilities to mere chance to create and to "improve" on,"organs of **such wonderful structure**, as the eye, of which we hardly as yet fully understand the **inimitable perfection**.*"* (Darwin.) Yet, since Darwin, science has become ever more materialist and atheist in its ways of looking at nature.

'Hence, the irrational insistence on blind luck explanations for everything —even for *Life,* even for *Mind,* even when it does not work. Yet, dare, as a scientist, academic, or public intellectual, today, to go against this dogmatic—no-intelligence-allowed—climate, *dare* to point out the obvious, that Living Nature is *saturated* with incredibly clever functioning, all the way up, and all the way down, and *you'll* be dismissed as a crank.

'Rather than admitting to the *amazing!, living, intelligence!* which keeps our own *stunningly, staggeringly!, stupendously, complex* bodies working, our hearts beating, our blood circulating, *right now!, this amazing second!,* and *every second!, 24/7,* we have, in the name of science, opted for a shockingly shallow and unrealistic idea of existence.'

'I think, Ollie, a lot of this is because we so hate to admit that we don't know. So, some of us take the old religious creation stories too literally, or, we resort to scientific sounding *'just so'* stories, like Darwin's, to try to explain things, then we hate to admit it when *that,* also, does not work.'

'Yes, I agree. Unfortunately so many now think we have a binary choice between (a) 'superstition' which they equate with all that is religious and spiritual, or (b) 'science' and 'rationality' which they equate with the materialist or atheistic belief system.

'Whether, or not, the *spiritual*, including intelligent creation, is a fact, is a valid scientific question, which we can try to answer. How accurate the world's religions and other spiritual traditions are in their depictions of the *super* natural and the *spiritual,* if these are real, is a separate question.

'Do you mean, on questions like whether god is a *single* force, as some hold, or a *hierarchy* of spiritual forces? Whether God is *One,* as in Islam, *Three,* as in Christianity, or One (Brahman), Three (Brahma, Vishnu, Shiva), *and* Many (other gods) as in Hinduism, or no 'god' at all as in Buddhism?'

'Yes. Or, for example, and of more immediate interest to most of us, **what happens when we die**? Are any of the traditional religious depictions of the afterlife, with horrible hells, that only sadistic gods [made in our image?] could devise, or boringly pious heavens correct? Is there any truth to the belief in human to human *reincarnation,* after a spell in the *spiritual* realms? Is *spirit perfect,* as some say, or is *spirit* itself evolving? Creation, even if intelligent, does not *seem* to be perfect, not, at least, from our human perspectives…

'Personally, Ali, I think that while the religions have got the *basic* idea of the *super* natural and the *spiritual* right, I think they are, all of them, probably inaccurate on many of the details. This is where the modern research into the *super* natural comes in, of which, unknown to many, there has been a lot, since the 1840s. We don't hear much about this modern research because it contradicts the materialist belief system which is so fashionable right now. But this extensive research provides overwhelming evidence for the *spiritual.*'

'By the way, Ollie, another contradiction in the totally materialistic approach, in science today, is, how can something, which—allegedly—has no intelligence or **purpose** to it, be "<u>**improved**</u>," as Darwin put it? Because 'improvement' implies existing function doesn't it? Yet as soon as a *Cell, a DNA, a Heart, or Liver* **functions** or works, it *must* have **purpose** to it—and reason behind it—otherwise it would not *work, or function,* at all.'

'Do you mean that without purposeful forces at work in *Living* things, nothing would ever work? Forces which point to intelligence…'

'Yes, *imagining, desiring, creating,* all imply *subjective Being or Beings* capable of imagining (creativity), that which is desired (love), and manifesting (creating): which is intelligence, love, and will, in action.

'Intelligent mechanisms require *intelligence(s)* to imagine [mind], to desire [love], and to create them [will]. The idea that Nature's *amazing, Soulful,* expressions are not intelligent or purposeful, is, so clearly, untrue.

'So much, now, is topsy-turvy, back-to-front and up-side down. Lies are truth and truth is a lie. I used to think, Ollie, that science was the most rational of activities. But when it comes to some of the *most important* things we can think about, including our very origins, it feeds us nonsense.

'I agree, it is sad. Perhaps our great scientific and technical successes, over the last few centuries, have, in some ways, made us lose our minds: ego, vanity, hubris: 'Those whom the gods would destroy they first make mad.'

'Another of the downsides to Darwinism, I began to notice, when I began to think a bit more, Ali, was that it has nothing useful to say about the *inner Being, the Mind and Soul,* the subjectivity and sentiency in living things. Nothing about the hard problem of *C O N S C I O U S N E S S.*

'Darwin's theory relates purely to the FORM or 'TAILS' sides of nature's holistic coin, to the MATTER and MECHANISM aspects only. It is solely about Nature's OUTER forms. Nothing about her *inner or 'Heads' sides,* her *Subjective* elements. Nothing about *Soul* or *Intelligence* in Nature. This was odd. Because, Darwin's *ideas* emerged, from his *intelligent Mind,* and his *curious Soul* trying, very *purposefully,* to make sense of nature, without, he argued, there needing to be any purpose, or *Soul,* or intelligence in nature.

69

'Which is self-contradictory…'

'Yes, yet, like many intellectuals of more recent times, he left *himself,* his *intelligence,* and his *purposes,* out of nature. He looked at nature, primarily, as all OBJECT, all 'BODY' and, literally, no *Soul,* and no *Spirit.* No *inner* aspects—all very LINEAR, all very DEAD, and MECHANICAL.

'Which is what 'MATERIALISM' is Ollie, isn't it? There can be no *Soul or Spirit or Anima* in materialism's dead view. There is just dead matter . . . which only seems to be *ensouled and alive . . .*'

'Yes, our *Faustian obsession* with the purely physical aspects of Nature, seeing her as *all beautiful Body,* but no *Spirit or Soul,* while it has brought us many material benefits, many luxuries and indulgences, has, tragically, totally cut us off from the true *Life and Soul, the true meanings,* and the *deeper meanings* of things.

'Thankfully, some of the world's earlier peoples still remember, for the rest of us. For them, all of reality, from rocks to plants, animals to planets, planets to stars, even the very galaxies, have *inner,* subjective *Soul and Spirit* aspects. For them, the entire universe, at all levels and dimensions, atom-to-mineral-to-cell-to-star, all the way up and all the way down, is an expression of $Spirit\ and\ Consciousness$, not 'DEAD MATTER.'

'Interestingly, the *Yogi-Adepts* of India, who are old-earth creationists in their spiritual vision, maintain that, over the eons, Life on Earth has waxed and waned across vast Life cycles, called *Yugas,* some of which *Life Waves* may correspond to the mysterious emergences, without plausible shared ancestors, of new species, as seen in the fossils.[45]

'According to the yogis, *Life's* history is not just an unfoldment of OUTER, MECHANICAL FORMS, and nothing more, but a meaningful unfoldment of *sentient interiorities,* of diverse *Soul* states for the different kinds of beings. For example, the fascinating world of the dinosaurs had its *own, mysterious, Soul feel.* Even now, we can still sense it… the amazing dinosaurs! It mysteriously arose, it served it's mysterious purpose, and it went.'

We continued on our way.

5 Life's Mysterious Explosion

'Religious creation stories, nowadays most people can agree, are not literally true, nor, really, were they ever *intended* to be. They simply symbolized to us that the ultimate *Sources of Life and Mind* were *intelligent,* not random.

'Most people have surely always realized that *'Spirit'* is an Intelligent Creative Energy, *which **"just Is,"*** not a childish, Father Christmas type god. No, and rather, *It/S/He, or Ish,* is an amazing, *all-pervading,* **Life-Source-Force-Energy-Being-ness,** which we can relate to in many ways, intellectual, emotional, personal, impersonal, with private or public ritual, inner or outer worship, prayer, meditation, some of us even arriving at powerful *spiritual* awakenings of various kinds, including self-realization and enlightenment.'

'A key point, though, for our *scientific* understanding, is that this *universal,* **Life-Source-Force-Energy-Being-ness,** not only *gives rise* to everything, but it is, literally, en-***Live**-*en-ing everything, *right now!, this amazing second!, and every amazing second!,* you, me, everything we see.

'So, if Darwin, and other thinkers of his time, had followed this more *sophisticated* understanding of *Spirit,* than their straw-man, god-with-a-beard, which they so mocked and disliked, (their own, too distant, Victorian fathers?), they would have been able to remain open to a wider range of creative possibilities, for the origins and history of Life.

'While it is true that Darwin's idea of **common descent** did seem able to account for *some* of the observed data, there was so much more it could *not* account for. This is why he *had* to ignore so much, in the natural worlds—and in the human *mental* worlds of evidence and classical logic—to try to make his theory seem true, even, as it became ever clearer that it was ***not*** true.

'If we rule out an *instant* creation—young-earth or old-earth—we have to opt for either (a) an ***intelligently driven*** evolution of some kind or (b) a MINDLESSLY driven evolution. Darwin opted for (b). If he had opted for an *intelligently* driven process his ideas would have made more sense.'

'Ok...'

'Because, if he had allowed for *intelligent causation* to be a *'natural'* part of reality, he could, also, have allowed for the real-world phenomenon of *sudden appearance*—vs. gradual. According to his theory, evolution *had* to be gradual. It could *not* be jumping. He insisted on this. But, the *sudden appearance* of new species, without plausible ancestors, is the rule, *not* the slow, progressive, unfoldment he predicted . . .'

'It is ironic, too, Ollie, that our science accepts the astonishing phenomenon of the *sudden appearance* of an entire universe, the Big Bang, but it does not allow that lesser 'Big Bangs,' of various kinds, including in biology, may, also, be possible, which is what the the fossils point to, isn't it?

'Although, even if Darwin *had* argued for an **intelligent** evolution, vs. rAndom, his theory of universally shared descent, from just one, or two microbes, in a branching, tree-like way would, *still,* be wrong, wouldn't it?'

'Yes, because even if *Life* arises *intelligently,* it clearly did not unfold from just one (or two) microbe(s). Rather, it arose in different ways—at different times—and in different places. ***That,*** would have been **accurate**.

'However, as Darwin did not believe in *intelligent* causation, his idea of a gradually branching tree of life, from microbes, followed by a long, but mindless evolution: microbe to worm, worm to fish, fish to amphibian, amphibian to dinosaur, dinosaur to mammal, ape to human, was not without logic. However: WAS IT TRUE? Logic, by itself, is no guarantee of truth.

'What *shatters* Darwin's idea, his **no-intelligence-needed** hypothesis of Life, is (a) that it depends on *blind chance* for all its creativity, which we, unlike he, know is *scientifically impossible.* (b) It makes no allowance for the *stunning complexities* of living things, nor, indeed, for the *second by second, intelligence* needed to operate living things. Nor (c) has it *anything useful* to say about CONSCIOUSNESS, MIND, and FEELINGS. Lastly, to keep itself going, this arid, sterile philosophy has to **ignore, discount, and dismiss**, more and more data which shows Darwin's ideas—and materialism in general—to be false. In the meantime, the pro-Darwin evolution community keeps very quiet about all this, especially when facing us, the general public. For example, how many people now realize, Ali, not only that natural selection is *vastly* less powerful than Darwin imagined—because it can only *refine* what already exists, it cannot truly create—but that *Life on Earth **did not** unfold, at all,* in the manner he predicted? It really didn't.'

'Go on...'

'Well, microbial life, it is currently believed, dates back 3.8 billion years. Yet, during the next 3.2 billion years, there is *no sign* of Darwin's imagined process—of descent with huge modifications—of new plants and creatures slowly emerging and branching out from those ancient microbes.

'Then, around 600–500 million years ago, the mysterious *biological* Big Bang known as the Cambrian Explosion took place, and *dozens* of new multi-cellular animal body plans appeared in a short period of geological time, including examples of all the basic animal body plans we know today.

'Yet, in contradiction of Darwin, there are no plausible fossil ancestors for these new forms in the *pre* Cambrian rocks. There are *no* signs that the many weird and wonderful creatures of the Cambrian Explosion, like Waptia, Marrella, and Canadaspis, descended, in any way, shape, or form, from the single-cell microbes of the previous 3.2 billion years.'

'So there's nothing 'evolutionary' about the Cambrian Explosion?'

'No, it's *revolutionary.* The Cambrian fossils show the geologically sudden appearances of many new Life Forms—*without* ancestor fossils.

'In ***The Naked Emperor: Darwinism Exposed***, Dr. A. Latham discusses how even world famous fossil authority S. J. Gould felt compelled to point out that Darwin's predictions did not match the data. [46] Gould wrote:

> 'Darwin invoked his standard argument to resolve this uncomfortable problem: the fossil record is so imperfect that we do not have the evidence for most of life's history. But even Darwin acknowledged that **his favorite ploy** was wearing a bit thin in this case. His argument could easily account for a missing stage in a single lineage, but **could the agencies of imperfection** *really* **obliterate absolutely all evidence** for positively every creature during most of life's history? Darwin admitted: 'the case at present must remain INEXPLICABLE; and MAY TRULY BE URGED AS A VALID ARGUMENT AGAINST MY VIEWS here entertained.' *Origin of Species,* 1859. [My emphasis.]

'Latham continues, "As already described, we do know of fossils before the explosion but these *completely overturn* the Darwinian model. For 3.2 billion years we have only microbial life... Then, in a period of about 50 million years: the Ediacaran Fauna which *appear unrelated* to the Cambrian fauna followed by the host of body plans of the Cambrian explosion."

'Latham quotes again from S. J. Gould's *A Wonderful Life:*

> "Thus, **instead** of Darwin's gradual rise to mounting complexity... Nearly **2.5 billion years** of prokaryotic [**non-nucleated**, bacterial] cells and nothing else, **2/3rds of life's history in stasis** at the lowest level of recorded complexity.
>
> "Another 700 million years of the larger and much more intricate eukaryotic [nucleated] cells [like algae] **but no aggregation to multicellular animal life.** Then in the 100 million year wink of a geological eye, three outstandingly different faunas – from Ediacaran, to Tommotion, to Burgess. Since then... [in complete conflict with Darwin] **not a single new phylum** or basic anatomical design [has ever arisen.]⁴⁷

'Gould continues, "you may *want* to read this sequence as a tale of predictable progress," *non-nucleated* bacterial cells first, then *nucleated* cells, then multicellular life, like worms, trilobites, snails, spiders, fish etc. But, he warns, *"scrutinize the particulars,"* and *"the comfortable story collapses."*⁴⁸ Why, for example, did life remain at stage one, for *2/3* of its history, if increasing complexity offers such great *biological* benefits?

'Why is there NO EVIDENCE AT ALL, he asks, for a *"slow and continuous rise of complexity,"* microbes to worms, worms to fish, fish to land-walkers, dinosaurs to mammals, mammals to apes, apes to humans?'

'So, for a vast time, **billions of years,** single-cells did *not* give birth to multi-cellular creatures, like worms, spiders or fish, as Darwin claimed?'

'Correct. Then, like everything else in existence—*as mysteriously as the arising of the amazing atoms and the magical stars*—biology's most dramatic *'big bang,'* the Cambrian Explosion, occurred, and many new animal body-plans appeared within a relatively short span of time of each other. However, as Gould mentions, and Michael Denton notes, there is ***no evidence*** for any kind of evolutionary arc[49] from single cell microbes to far more complex multi-celled creatures, like snails, worms, or fish—as predicted by Darwin.

'Then, since the ***not-tree-like-at-all*** Cambrian explosion, and in yet more dramatic conflict with Darwin's predictions, ***not a single*** new basic body plan has arisen. In fact, there are *fewer* body plans around than before, not more. Yet Darwin predicted that the new body plans would emerge out of the earlier ones. Except, this did not happen.

'Body plan?'

'It means basic anatomical design. There are over thirty, including: *Cnidaria,* sea anemones, corals, jellyfish. *Mollusca,* snails, slugs, mussels, octopuses. *Annelida,* marine worms, earth worms. *Arthropoda,* around 3.7 million species, including the classes *Arachnida,* spiders, mites, scorpions.

'*Insecta,* 1.5 to 3 million species, including ants, bees, beetles, flies and butterflies. *Crustacea,* shrimps, lobsters, crabs. *Platyhelminthes,* flatworms, parasitic flukes, tapeworms. *Nematoda,* round worms, over 80,000 species.

'*Echinodermata,* including starfish, sea urchins. Lastly, *Chordata,* the basic anatomical design to which vertebrates belong. 60,000 species including fish, amphibians, reptiles, birds, mammals and humans.

'The key point is that the ABRUPT arrival of so many new basic animal designs in the Cambrian FALSIFIES DARWIN'S ENTIRE THEORY!

'This massive problem is addressed in Stephen Meyer's, ***Darwin's Doubt: The Explosive Origin of Animal Life and the Case for Intelligent Design.***[50] The doubt in this title refers to Darwin's own doubts about his theory and his recognition that the animals of the Cambrian Explosion had no discernible ancestors. Darwin wrote:

> 'The difficulty of understanding the **absence** of vast piles of fossiliferous strata, which **on my theory** were no doubt somewhere accumulated before the Silurian (Cambrian) epoch is very great. I allude to the manner in which numbers of species of the same group **suddenly appear** in the lowest known fossiliferous rocks.'[51]

'Meyer explains how the sudden appearance of the major animal groups in the Cambrian is as discontinuous or jumping as the origin of Life itself. Darwin, though, very firmly believed that 'nature does not make 'jumps."

'Well, we can believe what we like, can't we Ollie? But it doesn't make our beliefs true, does it? Some people profess to believe, even today, that the Earth is flat, but that doesn't make it true. It seems that Darwin, and his later followers, **instead of following the evidence**, have, all too often, ignored it…'

'Yes, because, they, like he, think his theory is just too 'good' not to be true! For example, in their endless efforts to discount any data which contradicts Darwin, some of his modern supporters argue that the sudden arrival of so many new species in the Cambrian only *seems* discontinuous, because any precursors to the Cambrian creatures *may* have been ***soft-bodied*** and thus lacking in bones or shells which could be preserved in fossil form.

'Meyer notes, however, that plenty of soft-bodied organisms are preserved in the rocks from *before,* during, and after the Cambrian period.'

'Meyer also explores why random genetic mutations, held to be the very engine of Darwinian evolution by Darwin's modern supporters, he did not know about genes, **do not have the vast creative powers needed** to produce these new forms. Casey Luskin writes,

> "[Dr Meyer] cites multiple peer-reviewed studies showing that MULTIPLE COORDINATED MUTATIONS would be necessary to produce functional proteins, but these COULD NOT [POSSIBLY] ARISE within [the relatively limited] waiting times allowed by the fossil record. ... [This is] devastating to [all] Darwinian [style] explanations of the origin of complex features." [52]

'Dr Meyer, Luskin concludes, establishes intelligence—the mysterious, non-physical, *spiritual* quality, which we all know, and cannot live without, "as the only known cause capable of generating the **complex instructional information** and top-down design that are required to build the [new] animal body plans which appear so suddenly in the Cambrian period."[53]

'You know, Ali, after I began to have my own doubts about Darwin, and to do more research, I was shocked to discover just how much his theory, which we are all supposed to believe in—and be thought foolish if we do not —CONFLICTS WITH THE MOST BASIC FACTS OF NATURAL HISTORY.

'Crucially, that the species arose suddenly, *not* gradually. Darwin argued that the first body plan, say a single-cell alga or bacterium, could lead to the next body plan, say an **invertebrate** worm, which could lead to the next body plan, say a snail or a **vertebrate** fish, and so on. The FOSSILS REVEAL THE OPPOSITE. The basic body plans arrived in one go.'

'But Darwin took no notice?'

'Well, he dismissed the Cambrian data by claiming that the fossil records were very incomplete. There was just too little fossil evidence…

'But, as Sunderland says, our picture of Life's history is now far more detailed than when Darwin first published. Our museums are now "filled with over 100 million fossils…" Yet, do the fossils reveal "a continuous progression connecting all organisms to a common ancestor?" No, they don't.

'As Shedinger recounts in ***Darwin's Bluff***, Darwin really disliked admitting that a scientific hypothesis he was passionate about, might be wrong, even when well qualified colleagues were rightly skeptical. So, one of his regular tactics was to claim that the *data* was wrong, not his theory.'

'So he could be stubborn about his ideas?'

'Yes. His desire to be 'right' overrode, at times, his ability to stay close to the data—and to **follow the evidence**. For example, Art Battson observes that, after their very abrupt, their very non-Darwinian, appearances, the subsequent **stability** of the species through time "suggests the existence of natural processes which [**actively**] **prevent major change**."[54]

'So, he suggests, Darwin should have *inquired* into the fact that the species *do not* gradually evolve into new forms, as he at first thought, but remain *extremely stable*. Darwin should have tried to *explain* this, rather than clinging to his own hunches, just because he dreamed that—with his twin ideas of (a) ***"descent with modification"*** and (b) ***"natural selection"***—he might have discovered new principles almost akin to the laws of gravity.

'We, though, Ali, unlike Darwin, now know that cells use very **clever means** to ensure that they, precisely DO NOT RANDOMLY MUTATE in the way his theory requires. *Cells* work, very actively in fact, to PREVENT exactly the kinds of random genetic changes which Darwin's modern supporters say are the very sources of Darwin's alleged evolutionary process.[55]

'For 150 years it has been claimed that Darwin successfully explained the origins of the species, but that, sadly, there is just not enough fossil evidence to confirm his theory… Battson writes that this, strangely, "puts Darwin's theory in the curious position of explaining the data **we *don't* have** while ignoring the data we do," ⁵⁶ the tremendous STABILITY of the basic body plans and the stark fact that "NO TRANSITIONAL SPECIES [extinct or extant] showing evolution in progress between two stages HAS EVER BEEN FOUND."⁵⁷

'So, given, Ollie, that the Cambrian data contradicts Darwin, and that, despite all the searching, **no convincing transitional species**, fossilized, or living, have ever been found, why don't scientists develop theories to explain the **sudden appearances** and the **natural limits** to biological change,⁵⁸ rather than keep trying to force the data to fit Darwin's always unproven ideas? I thought scientists were supposed to *follow* the data, not to ignore it.'

'Because, if it is conceded that the basic animal body designs **arrived in one go,** and that the subsequent coming and going of the species is similar to the way our own designs appear, each new model abruptly appearing, stably lasting, without morphing into anything else, this points, far more logically, to *super* natural causation—than to blind chance. But this classical idea goes against Darwin and most of his followers materialist (atheistic) beliefs.

'Darwin predicted slow, *non-jumping* change, but, as Latham writes,

> 'Check any textbook on the subject and you will see that SUDDEN APPEARANCE AND DISCONTINUITY is the norm. The problem that Darwin so clearly saw, has not gone away with time, and the FOSSIL RECORD seems now to REFUTE the entire basis of his theory of [very slow] gradual changes.'⁵⁹

'Battson reflects on Darwin's speculations, which, despite all the propaganda, remain unproven to this day. The theory, it seems, is just 'too important' not to be true—even if, in reality, it is not true! He writes,

> 'Darwin's general theory of evolution may, in the final analysis, be little more than an *unwarranted extrapolation* from microevolution [minor variations] based more upon [materialist] philosophy than fact. The problem is that Darwinism CONTINUES TO DISTORT NATURAL SCIENCE.' ⁶⁰

'In the first edition of *On the Origin of Species* Darwin wrote,

'In North America the black bear was seen by Hearne swimming for hours with **widely open mouth**, thus catching, like a whale, insects in the water. Even in so extreme a case as this, ... **I can see no difficulty** [*are you joking?*] in a race of bears being rendered, by natural selection, more and more aquatic in their structure and habits, with **larger and larger mouthS**, till a creature was produced as monstrous as a whale.' [61] [See cover.]

'The fossils record no such fantastical transformations. No. New species appear **abruptly**—without plausible ancestors. They **stably** endure, without changing into anything else, and, eventually, they disappear into extinction.

'Then, the next Life Wave of similar or somewhat different species appears in the *same,* **jumping** way. It too is stable. It also does *not* give birth to major new forms, and, in time, it ends.

'Darwin was aware that the fossils *did* not, at the time he published in 1859, bear out his theory. He just hoped that, eventually, he would be proved right, and that the millions of fossilized species, gradually turning from one kind into different kinds—needed by his theory—would be found. They never have. Acknowledging this, Darwin asked,

'Firstly, *why,* if species have descended from other species by insensibly fine gradations, do we not **everywhere** see innumerable transitional forms? [*"everywhere?!"* Come on Darwin, NOWHERE *do we see ANY transitional forms,* let alone 'innumerable'!] *Why* [also] is not all nature *in confusion* instead of the species being, as we see them, well defined? [Separate and in discrete kinds.] [62]

'He was right to have doubts because his idea that **existing** species might morph into others—dead CHEMICALS to *Living Cells!,* single, asexual cells to multi-cellular worms, worms to fish, fish to dinosaurs, bears to whales—has never been supported. It was a linear, mechanical idea. We know how a river valley evolves. Keep pouring water down it, and it gets its shape.

'Darwin wanted to apply this same kind of linear, mechanical thinking, which worked well enough in geology, to very Life itself. But, he had *no evidence* that it *could* happen, nor, indeed, that it *did* happen. One thing we can say, though, is that like a lot of the things we make: "Many living things appear to be connected by design." For example, the shared skeletal patterns we see in vertebrates. The many shared *instincts* and *soul* qualities.

'Clearly, our Cars are connected, as we dig down through the layers of *'fossil'* vehicles in a junk yard—modern Fords all the way back to Ford Model T. The vehicles share many similarities and there *is* a succession. But, it is a succession driven by *ideas* and *intelligence,* not by mere *chance.'*

'But don't the fossils show some continuities of form?'

'Yes. But continuities of form, each new model, discrete from the previous, are not evidence for *'unintelligent'* design Many of our designs, each produced, by means of our own *super* natural, *almost* god-like input, show continuities of form, Cars, for example, or TVs and PCs.

'In ***The Naked Emperor: Darwinism Exposed,*** Dr. Latham writes:

> 'By looking in detail at the known fossils of the earliest tetrapods we can definitely see continuity between the lobe-finned precursors and those first land walkers. Any creationist must take this into account. When we look at either Acanthostega or Ichthyostega we see . . . they retain a sort of fish shape . . .
>
> 'We see here that, even though there is an unexplained L E A P to terrestrial locomotion, there is linkage with fish . . . such continuity is common in different groups of animals. This does not validate Darwinism, however. These are *not "transitional"* in the sense that Darwin meant.
>
> 'We see **too much that is suddenly new** to call the first tetrapods [four leggeds] transitional. There is **no gradual evolution** here but we do see earlier forms being a sort of template for the creation of later forms. The appearance of tetrapods is just one of the **saltations** [L E A P S] that characterize the fossil record.'[63]

'Darwin said evolution did ***not*** make L E A P S. It *had* to be gradual, to be true. Except, the linking fossils called for by his theory are *never* to be found—let alone "inconceivably great" numbers of them.' Latham continues,

'I hope that I will have demonstrated that there are indeed totally unexplained yawning **g a p s** in the fossil record but that there is also continuity between succeeding forms, hence the fishlike characteristics of the first tetrapods. They appear with all the attributes of land walkers, **suddenly** – but retain signs of their lineage. Darwinism requires smooth continuity always. We do not see this. There have been enough fossil beds examined (particularly in Greenland) for a clear picture of the fish to tetrapod evidence. The transition to tetrapod [four legs] is sudden. Darwinists will put this down to inadequate fossil preservation but this argument is now **wearing thin**.

'The challenge to the Darwinist is to SHOW US EVEN ONE CASE in the history of life where a macro change has occurred smoothly with well-defined, gradually transitional intermediate forms. DARWIN WAITED AND WE STILL WAIT.'[64]

'In ***Darwin's Bluff, The Mystery of the Book Darwin Never Finished***, Shedinger reminds us how Darwin promised, *repeatedly,* that *On the Origin of Species,* was ***just an introduction*** to his ideas, which he said he would later substantiate in a much larger, more authoritative book, backed up, "**in detail**," he said, with "ALL THE FACTS," with "references."[65] Yet . . . it never came.

'Darwin predicted *countless linking* species on their way to becoming something else, but NO CONVINCING LINKING FORMS, fossil or living, are ever to be found. *Sudden emergence,* followed by STABILITY, is the rule.

'It is no surprise, then, that Darwin never published his promised larger work. When he realized that he could not *substantiate* his ideas regarding the powers of MINDLESS NATURAL SELECTION to create and to evolve all life —he quietly shelved it. Yet, despite Darwin's *own* doubts, despite the disconfirming *data,* so many thinkers still seem to think it's their job to try to make Darwin *'right,'* while ignoring all the data showing that he was *not...*'

'But weren't Darwin's ideas, even if mistaken, an improvement on trying to treat the old religious creation stories as if they were science?'

'Yes, but religion isn't science, Ali, nor is it *meant* to be. It is about our *inner* relationships to the *deeper, Living, spiritual* sources of things. Our word, religion comes from the latin, *'re-ligio,' to 'tie' or to 'link'* back, to the *inner, Spiritual* sources of *Life*. It is not about, nor is it meant to be about, outer PHYSICS or CHEMISTRY.

'Darwin's idea that *mere chance* could evolve all Life, was always dubious, but it has morphed from a scientific hypothesis, which could be questioned, debated, tested and, if appropriate, *falsified,* to a secular, *pseudo-religion,* providing 'answers' which can no longer be questioned.

'But that's not 'science,' Ollie. It's politics and ideology, isn't it?'

'Yes, of course it is. As Battson says, Darwinism is now *distorting* science. Researchers now realize that it would take, not just a few, but **millions** of (rare!) random genetic mutations, **which cells very carefully protect against!**, "for dinosaurs to evolve into birds, flat plants into trees, fish into amphibians."[66] And as Geoffrey Simmons notes,

> 'Darwin wrote that whales came about as a result of **bears going to sea**. [Yet no one has a clue] how blow holes came about… or how the **internal lungs became connected** up to these holes in a way that prevents drowning. Or, how a **massive communication center,** found in their heads, came about. Or, how the ability to depressurize body segments during deep dives evolved. …
>
> 'Whales are not the only misfit to smooth [non-jumping] transitions … [Others] are kangaroos, **woodpeckers,** platypuses, **giraffes,** butterflies, **octopuses,** skunks, bombardier beetles, the red tide, **dolphins,** [sharks], **fireflies, tardigrades,** sloths, and **all microorganisms**. Maybe viruses, too…
>
> 'It's true, natural selection happens in a variety of situations, but IT DOESN'T CHANGE ONE SPECIES INTO ANOTHER.'[67]

We continued on our way.

6 True Nature

'For Darwin, although Life might look intelligent, and designed, a fact which he readily admitted, it was an illusion. Whether the laws of chemistry had an ultimate, *deistic-type* intelligence behind them was irrelevant, the laws of chemistry 'just were,' and these, plus enough 'lucky' chemical coincidences had led, he tended to think, to all Life on earth.

'There was, therefore, no need to refer to the *super* natural when trying to explain living things, anymore than we need to explain the formation of a river valley by appealing to the actions of some now invisible giant's hand.'

'But the problem, Ollie, is that, while (apparently) mindless natural laws can explain the random evolution of a landscape, can they account for Living things, all **far more complex** than anything we make?'

'Yes, because every gadget, Car, PC, or TV we create has to be *first* conceived in a Mind (its idea), for a purpose (its reason), its parts created from raw materials, cleverly extracted by us from the earth, and put together in very **specific** (non random) sequences to become a working mechanism.

'So, for the classical philosopher, Aristotle, all things, not only man made, could be thought of in terms of these four basic causes:

1. The **shaping idea** (Atom, Rock, Clock, Life Form): formal causation.
2. The **purpose** (what it's for, what it does): he called this final causation.
3. What it's **made of**: material causation.
4. How it's put together, **how it works:** efficient causation.

'And, until modern times, Aristotle's idea, simple as it might seem to us, was respected. Because, although simple, it made sense. Our Cars, TVs, and PCs are all imagined, mined, and refined out of the rocks and oils in the ground, and put together in **very specific sequences** of parts—by intelligent means. None of them 'fall together' by chance.

'Each embodies Aristotle's classic four causes: idea, purpose, materials and specific ordering to create viable mechanisms. The bodies of bacteria, worms, fish, birds, mammals and people are all forged from the raw materials of our planet into millions of highly purposeful, and totally specific sequences of parts by extremely intricate bio-chemical processes. Processes which make our most sophisticated technologies look basic.

'But our science today argues that all 'natural' things can be explained by reference to *just two* of Aristotle's famous four causes: (1) **'made of'** and (2) **'how it works**.' No need to consider formal and final causes anymore.'

'Yet, surely, all four of Aristotle's causes made sense. 'Made of' and 'how it works' do not account for ***'why'*** it does so . . .'

'Yes, but, it is now held that, either, there *are* no ***why*** causes, or, if there *are,* we do not need them to explain the existence of Living things. This is the approach, in modern science, known as method.o.logical materialism or **naturalism** (MM/MN). Two ways of saying the same thing. Where 'natural' means purely physical and no reference to the *super* natural is allowed.

'According to this MM/MN methodology, scientists must assume that all things have 'natural' causes, and nothing beyond everyday nature, like *Soul, or Spirit,* even if real, is needed to make sense of anything.

'While this can seem appealing, because we are born of nature, watered by her rivers and warmed by her Sun, it begs the question as to where **Nature herself** comes from? What is ***her*** true nature? Is she physical only?

'Or, is she *multi-dimensional?* Does she contain *Soul and Spirit* dimensions, *subtle, inner* dimensions, as the world's earlier peoples intuited, or is she **mono-dimensional,** as materialism claims?

'Since Darwin published, in 1859, we have often been told that: "Life is a purely *'natural'* phenomenon. When conditions are right, living things *naturally* arise, then, they *naturally* evolve. A progressive evolution towards ever greater complexity is *'natural,'* perhaps it is even inevitable. We do not need to resort to unscientific *fairy tales* about the *'Divine,'* or *'Spirit,'* or the *'super'* natural to explain anything." This approach was key to Darwin's theorizing. Perhaps there was a *'natural'* explanation for everything. Are not **nature**, and her laws, amazing? Why do we *need* to look deeper to explain?'

'Ok, Ollie, but the pleasing sounding word 'natural' when used to try to explain *all* things, even *Soul* qualities and abilities *like awareness, thinking, feeling, laughter, curiosity, creativity,* is question begging, isn't it? Because it does not answer these *more subtle* questions as to the *ultimate* nature of Nature herself or, indeed, of *Being-Life-Mind-Existence* itself.'

'I agree. Is Nature only physical? Or, does she consist, also, in *Soul* elements? Does she have *inner* dimensions, or is she OUTER only? Is she MINDLESS, or intelligent? Purposeless or purposeful?

'The **true nature** of ***All that Is***, of we, the amazing world around us, of the plants and the animals, of the planets, and stars beyond, lies at the heart of all our debates about Life's ultimate origins.

'While a purely materialistic approach can be used, well enough, to investigate what a *Star, Cell, Heart, Car or TV* is **made of**, and **how it works**, inquiries which our sciences are now very good at, it does not tell us **how** or **why** they arose in the first place, does it?

'I think that, for you, Ollie, *Cosmic Intelligence, Soul, Spirit, inhere* in Nature? They *pervade* her, don't they? They are not just beyond her?'

'Yes. Some call this way of understanding nature, *idealism,* or *pan-psychism,* where *all-pervading, Self-Existing Being,* [whatever we may label it, God, Brahman, Allah, Buddha Nature, the Tao, the name is unimportant], which **just Is**, not only gives rise to all things, but is ***within*** all things, enlivening all things.

'According to this understanding, *Spirit-Consciousness-Being* is not only *transcendent* to creation, but ***immanent*** to it, second by second, 24/7.

'The world's earlier peoples understood this, Ollie. They experienced the natural worlds as *en-Spirited and en-Souled . . .*'

'Yes, for them, all things had *Subjective, inner, Soul* dimensions. Unfortunately, despite all our modern cleverness, that *inborn,* intuitive sense of the *inner Livingness,* the *inner Soulfulness* of all things, has gradually faded, and it has been replaced by the dead-eyed 'god' of MATERIALISM. This is why our sciences cannot, now, allow for such possibilities. This is why, especially after Darwin, they assumed, more and more, that, either, *Soul and Spirit* do not exist, or that, if they *do,* it will, still, be possible to explain the living worlds *as if* they do not. This is why they say: "Regardless of whether or not this organism has intelligent sources, we can fully explain it in *'natural'* terms. It may *look* designed, but, in the right conditions, it could *'fall together'* by pure luck. There is no 'intelligence' or 'purpose' behind it."

'But, just as this approach won't work for a Car, PC or TV, it's unlikely to work for nature's organisms, all far more complex than anything we make.

'Equally, if we have *immaterial* Minds and Souls, which continue, after death, into dimensions beyond the physical, are we not part of an over-arching *Spiritual Nature* which is not only *pervasive* but *multi-dimensional?*

'This is why I prefer, Ali, to use the words materialist and MATERIALISM, over naturalist and NATURALISM, to describe the way of thinking which is currently considered scientific.'

'I guess 'naturalism' is ambiguous, because, while some see Nature as just PHYSICAL, others see her as *multi-dimensional,* with spiritual elements?'

'Yes. MATERIALISM is clear. It describes the belief, right or wrong, that only physical things, like CARBON, IRON, CO, AND H2O, are truly real.

'According to the thinking ruling science today, *Life and Mind* are not principles in themselves, but, rather, *Living Being, Awareness, Curiosity, Creativity, Love, Joy,* are all just accidental epiphenomena of a physical MATTER which is, itself, assumed to be MINDLESS and DEAD.

'I used to think like that. A way of thinking which seems ridiculous to me now. To conclude, just because the world's religious creation stories are not scientific descriptions of Life's genesis, not that they were ever intended to be, that Life is just a series of *implausible* chemical accidents—no, *impossible* is the better, more honest word—seems unhinged to me now.

'It is an *Emperor's New Clothes* way of thinking which, shockingly, far too many, in our otherwise very clever era, go along with. There is, clearly, far more at work in nature, than we can explain by referring to the PHYSICAL alone. Call it *Soul,* call it *Spirit,* call it the *Life Forces,* call it *Being,* the names don't matter. Our conclusions, though, do matter. Do we, and our Plant and Animal friends, live in an *amazing, Living, Cosmos of Intelligence and Meaning*—or in the MINDLESS UNIVERSE of materialist belief? Which makes more sense? Which is the more intellectually satisfying, overall?'

We continued on our way.

7 Fabulous Chance or Fat Chance?

'Although MATERIALISM is a *belief* system, not science, it gets a lot of its support from the scientific community, which, after Darwin, began, more and more, to consider *Life and Mind* to be purely physical things. Nothing *spiritual* to *Existence-Life-Mind-Being.* Materialism had, seemingly, triumphed. Religion and spirituality were just make believe.

'Many in science still accepted that the religions could bring us useful moral codes, but they considered the spiritually minded to be mistaken in seeing nature as *inherently* intelligent or purposeful in any way.

'In ***The Blind Watchmaker**,* Richard Dawkins famously affirmed that, 'Biology is "the study of complicated things that give the *appearance* of having been designed for a purpose," but, he quickly went on to say,

"the evidence of evolution reveals a universe *without* design."

'For Dawkins, like his hero, Charles Darwin, all things, even our *very minds,* can be explained by totally MINDLESS processes. This seems, not merely unlikely, but highly illogical to me, but, it is why Dawkins writes,

"Darwin made it possible to be an intellectually fulfilled atheist." [68]

'Darwin, he feels, showed us that materialism is scientifically justified. "Move on *spiritual* people, you're kidding yourselves." Yet, can life's intricacies, can our very *minds,* our very capacities for *knowing, thinking, feeling, and creating,* **really** be reduced to *mere accidents of chemistry?* Is this way of thinking honestly 'intellectually fulfilling'?'

'Dawkins clearly thinks so, doesn't he, Ollie? He's passionate about it.'

'Yes, but it doesn't *work.* It is *scientifically* hopeless. Interestingly, in a letter to the distinguished astronomer John Herschel, Darwin wrote,

> "One cannot look at this Universe with all [its] living productions & man without believing that **all has been intelligently**

designed; yet when I look to each individual organism, I can see **no evidence** of this." [69] [Ah, dear Darwin, you cannot have been looking very closely! Try making just one cell!]

'That's intriguing. Darwin was open to the classical idea of intelligent causation but then he reversed himself?'

'Well, he was conflicted. On the one hand, he could see powerful evidences for *intelligent and purposeful* functioning all around…'

'On the other, Ollie, it is not possible to see nature's *deeper* causes physically, and life can be hard to make sense of at times, can't it?'

'Yes, that's true, and it is, perhaps, materialism's most powerful argument. However, in his foreword to the 20th Anniversary edition of Philip Johnson's bestseller, ***Darwin On Trial***, Michael Behe notes that,

> "the situation for **materialistic** [**atheistic**] origin-of-life theories has gotten substantially worse [since Johnson first published]. Broadly speaking, for decades there have been **two categories of [materialist] origin-of-life theories**: [a] the "metabolism-first" view, where metabolic reactions in an enclosed space **precede** the occurrence of genetic material; and [b] the "genetics-first" view, where a DNA-like polymer that is **capable of carrying information** precedes cells.
>
> "The partisans of both camps have offered **devastating criticisms** of each others views, so that none are left standing… the **only reason** at present to believe in a materialistic [non-intelligent] origin of life is if one holds it as a postulate that life **must have had** a materialistic [or non-supernatural] origin." [70]

'Which is what scientists *are,* currently, taught to assume. So, some have argued that the *intricate* molecules used to carry Life's CODES might **naturally** arise where a *non-living* mixture of proteins and RNA molecules reached a point at which the mixture might start to self-replicate. This hypothesis is called RNA-world, origin-of-life thinking.'

'But why would such a stage arise, Ollie? Replication, by itself, doesn't mean something is *alive,* does it? Crystals *replicate,* but they are not alive as we are. And don't **RNA** molecules already contain *complex instructions* and in.FORM.ation, the origins of which remain unknown?'

'Yes. As William Dembski explains in *No Free Lunch: Why Specified Complexity Cannot Be Purchased Without Intelligence,* there is no such thing as *'free'* software. Yet: all living things are run by this, same, mysterious, **DNA CODED** software, which makes our most sophisticated IT look *basic.*

'The current, materialistic, claim is that the Sun's energy, which is *'free,'* could overcome the dis-organizing forces of entropy and, by chance, create life's **CODEs,** *Cells, Shells, Teeth!, Blood, Hearts!, Kidneys,* and so on.'

'But isn't that view as unrealistic as the idea that monkeys could randomly type some *'free'* Shakespeare, that a wild fire, with plenty of *'free' fuel,* could generate anything other than burnt wood, or that a tornado blowing through a junk yard could assemble *a 'free' passenger jet!'*

'Yes, of course it is. As Dembski and Witt say, the chances of a pre-life chemical soup 'randomly' burping up the very clever, *selectively* permeable, cell wall, which protects the DNA inside, at the *same* time as the DNA needed to CODE for that *same* cell wall, are far beyond improbable.

'They say it results in a number so high that: "it stretches the ability of biochemists to calculate it. The best they can do, is set a lowest possible figure, which *surpasses* by untold trillions of trillions of trillions of times the universal probability bound of 1 chance in 10 to 150," [71] mentioned earlier.

'Even Jacques Monod, himself a materialist, and originator of the famous phrase that just *"chance"* (chemical luck) + *"necessity"* (regular natural laws) could, he believed, explain everything in Nature, admitted that:

> 'we have no idea what the structure of a [hypothetical] primitive cell might have been. ... the **simplest cells** available to us for study have nothing **"primitive"** about them. ... The development of the metabolic system, which ... must have "learned" [how, he doesn't say!] to mobilize chemical potential and to synthesize the cellular components, [all by pure, aimless, chance!] poses Herculean problems. So also does the emergence of the *selectively* permeable membrane without which there can be no viable cell.

'But the major problem is the origin of the GENETIC CODE and its translation mechanism. ... The code is meaningless unless translated. The modern cell's translating machinery consists of at least fifty macromolecular components WHICH ARE THEMSELVES CODED IN DNA: THE CODE CANNOT BE TRANSLATED OTHERWISE THAN BY PRODUCTS OF TRANSLATION [emphasis in original]. It is the modern expression of omne vivum ex ovo [Life only comes from Life, so which came first, chicken or egg?].'[72]

'In *The Naked Emperor: Darwinism Exposed,* Dr. A. Latham writes,

'Far too often I hear experts **confidently** postulate about the first cell membranes. They know that **the first life** had to have a **protective coat [outer wall]** to hold the DNA and all the incredibly sophisticated molecular machinery inside the bacteria. ... [yet] the actual details of the [cell] membrane must be coded for in the DNA of the cell.'[73] [Behe refers to this as *irreducible complexity.*]

'As Latham the *spiritualist* and Monod the **materialist** both point out, *how could* the cell's very clever, DNA **coded** software arise, absent the protective cell wall, which the software *inside* the cell CODES for? The cell's wall and the DNA within, coding for that same wall, are *irreducibly* complex.

'They must arise together or not at all. Is this true? Yes. Does it point to blind chance? *Seriously?* Does it point to *inherent intelligence* behind living things? It makes sense. That was the classical, pre-Darwin hypothesis. But, if we continue to insist on the atheistic worldview we can never concede this.'

'So the battle, Ollie, is between the materialists, like Darwin, on the one hand, and, on the other, those giants of western thinking, like Plato, Galileo, Newton, and Planck, who were all *spiritualists*, who all believed that there is more to *Life,* and *Mind,* on our planet, than mere chance?'

'Yes. One of the great ironies of this debate, Ali, is that those who charge people, like you and I, with being naive and simple minded, because we cannot worship *blind chance,* as our supreme creative principle, never stop to consider **just how many utterly nonsensical things** *they* must believe, *"all before breakfast,"* as Lewis Carroll might have put it, to avoid the classical insight, that *Being and Beings—Cells to People, Atoms to Stars*—arise from *vastly intelligent, spiritual forces,* which *'just Are,'* rather than from the totally MINDLESS forces, which they worship, which, no less mysteriously, must *'just be.'* Which makes more sense, DUMB universe or *intelligent?*

'Interestingly… another astronomer, Sir Frederick Hoyle, wrote:

> **'Life cannot have had a random beginning…** The trouble is that there are about two thousand enzymes, and the chance of obtaining them all in a random trial is only one part in $10^{40,000}$ [1 followed by 40,000 zeros] an outrageously small probability that could not be faced **even if the whole universe consisted of organic soup**.[74]'

'Once we see, however, that the probability of life originating at random is **so utterly minuscule as to make it absurd**, it becomes sensible to think that the favorable properties of physics on which life depends are in every respect deliberate … . It is therefore almost inevitable that **our own measure of intelligence must reflect … higher intelligences … even to the limit of God** … such a theory is so obvious that one wonders why it is not widely accepted as being self-evident. **The reasons are psychological rather than scientific.**'[75]

'Later, in 1995, Eastman and Missler added to this debate that,

> 'Hoyle's calculations may seem impressive, but they **don't even begin to approximate the difficulty of the task**. He only calculated the probability of the spontaneous generation of the proteins in the cell. He did not calculate the chance formation of the DNA, RNA, nor the cell wall that [protects the cell].
>
> 'A more realistic estimate for spontaneous generation has been made by Harold Morowitz, a Yale University physicist. Morowitz **imagined a broth of living bacteria** that were super-heated so that all the complex chemicals were broken down into their basic building blocks. After cooling the mixture, he concluded that the **odds of a single bacterium re-assembling by chance** is 1 in $10^{100,000,000,000}$. This number is so large that it would require several thousand blank books just to write it out. To put this number into

perspective, it is MORE LIKELY that **you *and* your entire extended family** would win the state lottery EVERY WEEK, for a MILLION YEARS, **than for a bacterium to form by chance!** '76

'We need to take notice of these facts, not *ignore* them, because they contradict materialism. We need to challenge those who endlessly try to sell us the *unscientific idea* that: "Life is just a cosmic fluke which started by accident and evolved by accident. We ***'know'*** this to be true, and any dissenters from these truths are either ignorant or just being awkward!"

'This is what we are up against: a mentality that conflates science—the openminded exploration of *all* of reality—PHYSICAL and *Spiritual*—with MATERIALISM, the belief that unless we can physically see it, it cannot exist.

'Not noticing, all the while, that we cannot physically see our ideas, or our intelligences, or mathematics when held in our minds.

'We cannot physically see love, or compassion, or strength, yet they are all utterly real. This dreary, atheistic ideology which, dismissing the *spiritual* out of hand, insists that such stunningly complex things as *Living Cells!, or Hearts!, or Eyes!, or Brains!* could arise by pure dumb chance, when nothing could be further from the truth.

'Is this why we are asked to pretend, today, that a soaring bird, a swooping bat, or a graceful trapeze artist are all the purposeless and meaningless products of a wholly RANDOM process—and feel fulfilled?'

'But why,' Ali sighed, 'do more people, *especially academics*—who are meant to be the *brightest* amongst us—NOT notice that this is, *literally,* what all these beautiful things would be, if MATERIALISM were actually *'true?"*

'I don't know, Ali. All I can say is that it is a basic tenet of science today that THERE IS NO INTELLIGENCE OR TELEOLOGY (PURPOSE) IN NATURE. According to this strange view, no one's kind heart or skillful hands, no one's *Bright Mind or brilliant plans,* are anything more than accidents of a totally MINDLESS EVOLUTIONARY PROCESS. The same for our *Blood* and our *Brains,* for the DNA CODING in our *Cells,* and all the other amazing, *24/7,* mechanisms which make our *physical lives* possible. Any academics who dare to question Darwin's unreal paradigm of unintelligent design put their careers at risk. It is odd, too, that scientists today are thought to be clever enough to work out that nature is mindless and without *Soul or Spirit,* as thinkers like Darwin and Dawkins argue, but not clever enough to work out that this may not be true—quite the opposite in fact.

'That *Living Nature* is, as is obvious to most ordinary people, *pervaded by intelligence.* It is also sad that most of those who follow these arid, *soul and spirit* denying ways of looking at reality, refuse to take any interest in the extensive modern evidence for *mind-beyond-the-body* phenomena, like near death, out of body, and other spiritual experiences (NDEs, OBEs, OSEs).

'Evidences which amply falsify materialism. Their minds are closed. They don't to want to know, that their Darwinian 'emperor' has no clothes…

'I guess, Ollie, this is one reason why our current science culture lacks the intellectual curiosity to consider that a *spiritual* intelligence, or intelligences, might express creation in many different ways, not just through our *Souls,* but through the *Souls* and the senses of many other organisms, all with their own unique qualities of sentiency and awareness.

'Did you know that there are dogs who enjoy skate-boarding and parrots who like to dance, in time, to our music. You can see delightful clips of these online. Who are we to say that these amazing *Beings* have no *Soul* or reasons for existing, just because we don't know exactly what they're *'for'?'*

'Ah, my dear Ali, I do so agree. Michael Denton says, for us to build a scale model of **just**. . **one**. . **tiny**. . **cell,** at one tennis ball sized atom *per second* would take us hundreds of *thousands of years!* Yet, each cell, of which there are **trillions** in our bodies, achieves this at supercomputer speeds, *in just a few hours.* Darwin didn't know these things, but *we* do.

'How can anyone, who is actually thinking, believe that these kinds of things are good evidences for *purposelessness* or for *unintelligent* design?

'Don't like the traditional religions? It's understandable. Too often it was their way or the highway. But the idea that *Nature* contains no *intelligence* or purpose or *Soul* is not science, Ali, it's materialism *pretending* to be science.

'Yet, dare to point this out, and you're the crank…'

'Oh yes! Look, no one should have a problem with Darwin's idea of a universally shared descent, in a branching, tree-like pattern, chemicals-to-bacterium, bacterium to worm, worm to fish, fish to bear, bear to whale, shrew to bat, dinosaur to bird, mammals to apes, apes to humans, **if there was any good evidence for such a process**. Our only debate, then, would be as to whether it was a random process or one *intelligently* driven. But, as it is quite clear, by now, that *no such process,* as Darwin described it, took place at all —random *or* intelligently driven—shouldn't we move on?

'Darwin's three main ideas of (a) **common descent,** (b) **random variations** and (c) **natural selection** of random variations, were worth testing. They have, however, all been tested, very thoroughly by now, and they have all been falsified very thoroughly by now.

'John Lennox, an Oxford professor of math, notes: "(1) Life involves a complex DNA database of D I G I T A L information. (2) The only source we know of such language-like complexity is *intelligence.* (3) Theoretical computer science indicates that unguided chance [luck] and necessity [laws] are incapable of producing semiotic (language-like) complexity."

'Modern ID expert, Werner Gitt writes, "It has never been shown that a CODING SYSTEM and semantic [meaningful] information could originate by itself… The information theorems predict that this will **never be possible**. A purely material origin of life is [thus ruled out]." 77

'Lennox adds that, "one would have thought that scientists would prefer an explanation [intelligent causation] that explains a given phenomenon over an explanation that does not. The fact that this is not the case in thinking about the origins of life shows that **an a priori materialism** can produce a **profoundly anti-scientific attitude** – [and an] unwillingness to follow evidence where it clearly leads simply because one does not like the implications of so doing." 78

'He's right. Especially, as the fight here is *not* about practical, everyday science, which we can conduct without needing to think about Existence's *deeper* causes. For example, we know how to use electricity. Whether Life's origins are **intelligent** or RANDOM has no impact on this useful discovery.'

'So we could call this kind of science, **'practical or technical science?'**"

'Yes. But what we can refer to as **'philosophical science'** is mostly just about trying to make *sense* of things. Our discoveries in natural history, like the Cambrian Big Bang, the Dinosaurs, or our ideas on whether Darwin was right, or not, have little technical or practical impact. They are just ways of trying to make sense of our origins and of Life's history through time.

'Yet, while *practically not* important, the stakes are, ironically, much higher. Whether Life has *meaningful* causes, as most *classical* thinkers held, or random, as Darwin argued, has *huge implications* for us, for our relationships to each other, and for our relationships to Existence itself.

'Are our *spiritual* insights real or not? Are there *Higher Power(s),* we can *inwardly* turn to, or not? Are our *mystical* experiences, including ones *of oneness, of merging with the Whole, of realizing the Essential Goodness of Existence, reflective of a deeper, Spiritual reality, or not?*

'If materialism is true, the answer to all these questions has to be "no." If it is wrong, the answer is probably "yes." The implications are vast . . .'

'But given that so many now see *'science'* in opposition to all things *spiritual,* I don't see how this war between religion and science can end?'

'Well, these are not issues for practical, everyday science. The battles are *ideological.* The real fight is not, in any case, between religion and science, but between the **materialist** and the *spiritualist* world views. The materialist camp has sought, since Darwin, to persuade us that 'SCIENCE' belongs to its world view. It doesn't. Because true 'science' is but a series of methods for investigating reality, be reality purely PHYSICAL or be it, also, $spiritual$.

'We now know, unlike Darwin, that *Life could never* start by *chance,* nor, could it *evolve* by such means. Yet, for too long, those who question Darwin have been shouted down and cancelled, not because they are wrong —anymore than Galileo was wrong—but because they were in a minority.

'As far as the religions are concerned, while they can be unreasonable at times, their **basic claim** that Life's ultimate causes are ***intelligent,*** not dumb, is logical and reasonable. A similar realization, arrived at by modern, *non-religious* means, confirms their ancient insight. But that's as far as it goes.

'Modern observations of intelligence and design in Nature confirm *some* of humanity's ancient intuitions about Life's deeper sources, but we *cannot,* and we do not need to, go from reasoned, logical, admissions of intelligence and design in nature, to endorsing a particular religion, its beliefs about heaven and hell, its recommendations for salvation and so on.'

We continued walking.

8 Genetic Copying Errors

'If we remain open to the classical view that *Living Nature* includes subtle, *non-physical, spiritual aspects,* we can comfortably allow for intelligent causation. But if we rule this out, either as a matter of our methodology[79] or our philosophy, we will be forced to try to explain all elements of *Life,* even our very *Minds,* as though they are purely PHYSICAL things.'

'Yet, we know, very well by now—unless our philosophy does not allow us to know it—that mere chance can *never* create a microbe in the first place, let alone evolve it into a worm, snail or spider, and, later, beings with the minds and instincts of eagles, bears, or humans.

'We also know, very well by now that a species can only vary *within* but **not beyond the limits set by its gene pool**. There is *no data* for the *extreme levels of species variability* Darwin called for—with *huge* changes all along the way—microbes to worms to fish to dinosaurs to bears to whales.

'What about wolves to dogs? Isn't that an example of species variation?'

'Well, no dog, not even the little *Chihuahua,* is a **new species**. Such minor, *intra-species* variations cannot explain (a) how Darwin's hypothetical common ancestor microbe might arise in the *first place,* nor, (b) how it might turn into very different snails, whales, flowers, or people.

'Note, too, that the modern version of Darwin's theory—neo-Darwinism —which attributes such absurdly unrealistic powers to random genetic mutations, *cannot explain* where the original, non-mutated genes came from in the first place. It just *huffs* and *bluffs* at us that they must have *'evolved,'* somehow or other, and it hopes we are not paying attention.

'Because, even *'basic'* microbes are vastly more complex than Darwin and his steam age friends imagined. They could *never* arise by chance. We know this! This is resisted, in academia today, not because it isn't true, but because it contraries Darwin's fashionable, materialist (atheistic) worldview.

'For a *single-cell* microbe to turn into a *multi-celled* worm, spider, or fish, as called for by Darwin's ideas, **vast amounts** of *completely new* genetic information, far beyond the original cell's gene pool, would be needed. Darwin had no idea *how* the **incredibly vast variations** needed by his theory might arise, then, gradually, turn a microbe into a worm, crab, snail, octopus, fish, dinosaur, bird, bat, bear, whale, ape, and, finally, human.

'However, some time after Darwin published, Gregor Mendel's earlier work on genes was rediscovered. It showed how genetic variations *within* species arose, and how this shuffling of *existing* genes, within *existing* species, had been *intelligently* selected, by clever human breeders, in their pursuit of new or improved varieties of dog, pigeon, or tulip for example.

'In the wild, it was this *natural* shuffling of ***existing*** genes, in ***existing*** species, which favored the superior survival rate, or 'natural selection,' of those species members with the gene mix best suited to the prevailing conditions, for example, finches with bigger beaks, or the fittest bears.

'Mendel's discovery did not explain, though, how a microbe, worm, or fish, might, gradually, *transform* into an ***entirely other*** kind of life form, as predicted by Darwin: microbe-to-worm, worm-to-fish, fish-to-amphibian, amphibian-to-dinosaur, dinosaur-to-mammal, bear-to-whale, ape-to-human.

'Mendel had simply worked out how a species *existing* gene pool allowed for minor—*but limited*—variations *within* that species.[80]

'So Mendel did not discover that a species could, gradually, be changed —*by chance* or *by people*—into a completely new one?' Asked Alisha. 'Via, for example, a sea-going stage, like fish, then dinosaurs, then mammals *(chordates),* then mammal carnivores *(order),* to carnivore family *(canidae),* to genus *(canis),* to a species like wolves, *(canis lupus),* as Darwin predicted?'

'That's right. But, later, after Mendel, some people wondered whether random genetic mutations, or **RGM**s, which are genetic **copying mistakes**— the effects of which Mendel did not explore—might not generate *new* genetic information, and lead, eventually, to entirely new plants or animals, which, if true, might make Darwin's ideas hypothetically possible.

'This idea is called neo-Darwinism because Darwin did not know about genes. **RGM**s are, however, copying mistakes in the genetic code, which, we, now realize, all cells, very actively, protect against.[81] When **RGM**s do, occasionally, escape the Cell's extremely clever, self-correcting mechanisms, they are generally harmful, sometimes lethal, and no more do they provide organisms with useful new constructional information than a poorly copied microchip would be a useful new feature in a computer.

'In ***God's Undertaker: Has Science Buried God?*** Professor John Lennox tells us how the "incredibly precise duplication of DNA is not accomplished by the DNA alone: it depends on the presence of the living cell." John Lennox quotes biologist James Shapiro who writes,

> 'It has been a surprise to learn **[just] how thoroughly cells protect themselves against** precisely the kinds of accidental genetic change that, according to conventional [Darwinian] theory, are the [very] sources of evolutionary variability. By virtue of their **proofreading and repair systems**, living cells are not the passive victims of the random forces of chemistry and physics. **They devote large resources to suppressing random genetic variation** and have the capacity to set the level of background localized mutability **by adjusting the activity of their repair systems**.'[82]

'So it's unlikely that random genetic typos could drive the extremely powerful evolutionary processes Darwin had in mind?' Said Alisha.

'Yes. Yet, currently, any scientists pointing out that these very clever error correction processes in cells point, not to *'luck,'* but, to an amazing level of *intelligence* at work in living things risk being told that they are being fanciful and *unscientific!* [Ah, Dear Reader, I find it so absurd and embarrassing, of our current culture, that we need to point out all this, that should be utterly obvious!]

'The discovery of these sophisticated error correction processes *falsifies* Darwin's core theory of descent with vast, almost ***infinite*** *modifications:* bacterium to worm, microbe to fish, microbe to man, bacterium to oak, bacterium to bear, bear to whale, bacterium to fish to shrew to bat.

'Random **genetic typo mutations**, marketed to us, by Darwin's modern supporters, as the engine of a vastly progressive evolution—**particles to people**—cannot build the new, and any minor advantages, which they occasionally confer, almost always come at a cost.

'In *The Edge of Evolution: The Search for the Limits of Darwinism,*[83] biochemist Michael Behe explains that random genetic mutations are ***not*** constructive. For example, the well known *sickle cell* mutation is *not* desirable, but it does have the one benefit that it can confer some immunity to malaria. Generally, though, random genetic mutations lead to harms and to the **loss** of genetic information—not to construction.

'Although Darwin's theory can explain minor differences in organisms, like different varieties of finch, pigeon, or dog, RGMs could not drive the extremely powerful transformational processes he had in mind.

'Michael Behe adds that there is **no evidence** that the barriers, between the species, up to the level of families, could ever be breached by *any* random process. Despite their best efforts, clever, focused, human selectors have **never been able** to cross even the species barrier.

'Chihuahuas remain, in essence, *Wolves*. Bacteria are still bacteria. Tulips are still tulips. Pigeons are still pigeons, and fruit flies are still fruit flies.

'Dr. Durston, a scientist writing in *Evolution News* notes,

'There is mounting evidence that most, if not all the key predictions of the neo-Darwinian theory of macroevolution are being falsified by advances in science. ... Ask computer programmers what effect ongoing **random changes** in the Code would have on the integrity of a program, and they will universally agree that it **degrades the software**. ... deleterious mutations [do not create biological life, they destroy it].' [84]

'Genetic mutations are, almost always, *not* beneficial, and, far from causing a genetically deformed creature to be turned into a new life form, would lead to it being eaten. Even antibiotic resistance in bacteria is caused by rare, genetic copying mistakes, which are corrected out of the population when the antibiotic is no longer used.

'In *Darwin Devolves,* Behe notes that Darwin did *not show* that living mechanisms, be they microbes, or organs like livers, eyes, or brains, *really could* be built by natural selection. He just argued that it *'might'* be possible.

'Behe also points out that, while Sir Francis Bacon suggested, at the dawn of the modern era, that scientists need not worry about the ultimate *'purposes'* of things, be they Atoms, Rocks, or Horses, in order to do good practical science on them, to deny that anything in nature has *teleology,* or purpose, as scientists are currently required to do, is, so clearly, untrue.

'As Behe says, while we may not know what a Horse, Dog or Human is *'for,'* we *do* know that its heart and its lungs, its mouth and its limbs, all have *teleologies,* or purposefulnesses, which make the fascinating, living, mysteries which we call Horse, Dog, and Human, possible.

'Behe adds that random genetic mutations will be selected, and spread through a population, if they confer an immediate advantage. But, this *always* comes at the cost of the *loss* of genetic information.[85] RGMs do *not* build. They occasionally bestow minor benefits, but, generally, they degrade.

'So the idea that a dinosaur foreleg could gradually turn into a bird's wing, *with feathers,* due to a long series of *genetic typos*—which all cells actively *protect against*—is no longer good science?' Mused Alisha.

'To be honest, Ali, it was *never* good science. James Perloff, author of *Tornado in a Junkyard,* a title referring to astronomer Sir Frederick Hoyle's comment that it is as likely that Life could arise by chance as it is that a hurricane blowing through a junk yard could assemble a Jumbo Jet, writes,

> "According to Darwinism, single cells eventually evolved into invertebrates.., then successively into fish, amphibians, reptiles, and finally mammals. Darwin said this occurred from creatures adapting to environments. . . . [But Mendel's] discovery of genetics threatened this claim. **New organs require new genes**."

'However, as Perloff says, merely "moving into new environments doesn't give you new genes." This initially stumped Darwinists. Then, he says, they thought they had found a solution. "**Random mutations**—copying mistakes in the genetic code—**occur very rarely**, but [they] DO alter genetic information." So Darwin's modern followers said "animals gained new genes by chance mutations…which they adapted to evolve into higher forms."[86]

'However, Perloff notes, "chance mutations are to the genetic code **what typos** are to a book: they *remove* information, but do not improve it. In humans, mutations cause sickle cell anemia, cystic fibrosis, hemophilia, Down's syndrome, and *thousands* of other genetic diseases . . . even the rare "beneficial mutations" evolutionists trumpet – such as bacterial resistance to antibiotics – actually result from functional [genetic] losses."

'Take the *feathered* wings of Birds or the *featherless* wings of Bats. They appear in the fossil records **suddenly,** and **fully formed**, in the usual, **non-Darwinian**—and **non-evolutionary**—way. There is **no fossil data** to suggest that Bird or Bat wings v e r y s l o w l y evolved from wingless dinosaurs in the case of birds, or wingless mammals in the case of bats.

'As Michael Denton explains in *Evolution: A Theory in Crisis:*

> 'It is not easy to see how an impervious reptiles scale could [as Darwinians claim] be converted gradually into an impervious feather **without passing through a frayed scale intermediate** which would be weak, easily deformed and still quite permeable to air. It is true that basically the feather is indeed a frayed scale – a mass of keratin filaments – but the filaments are not a random tangle but are **ordered in an amazingly complex way** to achieve the tightly intertwined structure of the feather. Take away the exquisite co-adaptation of the components, take away the co-adaptation of the hooks and barbules, take away the precisely parallel arrangements of the barbs on the shaft and all that is left is a soft pliable structure **utterly unsuitable to form the basis** of the stiff impervious feather. The stiff impervious property of the feather which makes it so beautiful an adaptation for flight, depends basically on such a highly involved and unique system of co-adapted components that **it seems impossible that any transitional feather-like structure could** possess even to a slight degree the crucial properties. In the words of Barbara Stahl, in *Vertebrate History: Problems in Evolution,* as far as feathers are concerned, "**how they arose initially, presumably from reptiles scales, defies analysis.**" [87]

'But don't Darwin's supporters say that birds feathers arose from reptile scales because they are made of *similar* materials?'

'Well, we would not argue that airplanes evolved from cars just because they are made of similar materials…'

'Don't they also say that birds lungs evolved from the standard vertebrate lung? Because birds and dinosaurs do share some similarities...'

'Yes... but this is what biologist Michael Denton says about this:

> 'In all other vertebrates [including dinosaurs] the air is drawn into the lungs through a system of branching tubes which finally terminates in tiny air sacs, or alveoli, **so that during respiration the air is moved in and out through *the same* passage**. In the case of birds, however, the major bronchi break down into tiny tubes which permeate the lung tissue. These so-called para-bronchi eventually join up together again, **forming a true circulatory system so that air flows in one direction** through the lungs. This unidirectional flow is maintained during both inspiration and expiration by **a complex system of interconnected air sacs** in the bird's body which expand and contract in such a way so as to ensure a continuous delivery of air through the para bronchi. ... NO LUNG **in any other vertebrate species is known** WHICH IN ANY WAY approaches the avian system.
>
> 'Moreover, it is identical in all essential details in all birds as diverse as hummingbirds, ostriches and hawks.
>
> '**Just how such an** UTTERLY DIFFERENT respiratory system **could have evolved gradually** from the standard vertebrate design is fantastically difficult to envisage, especially bearing in mind that the maintenance of respiratory function is absolutely vital to the life of an organism to the extent that the slightest malfunction leads to DEATH WITHIN MINUTES.' [88]

'Perloff points out similar flaws in Darwinian thinking when it comes to our understanding of life at the cellular level. He writes that, for a Darwinian account of Life to be true,

> "the primordial cell must have perfected—in the span of one lifetime—the process of cellular **reproduction**; otherwise there never would have been a second cell." Yet, he says, despite the demonstrated I M P O S S I B I L I T Y of such a process, people are

still being "taught that life began [*not* due to an, intrinsic intelligence, which *just Is,* but] from a fortuitous arrangement of chemicals."⁸⁹ [which, very, very, very luckily, just happened!].

'Well, we are taught about Life's origins *as if* materialism is true, aren't we, Ollie? Which begs the big question: *Is* materialism *true?'*

'Yes. But Perloff isn't buying it. He says: "According to Darwinism, single cells eventually evolved into invertebrates (creatures *without* backbones, like worms, snails, jellyfish), then successively into [creatures *with* backbones, like] fish, amphibians, reptiles, and finally mammals. Darwin said this occurred from creatures adapting to environments."

'But, Perloff asks, if single-cell *"bacteria* evolved successively into *invertebrates,"* then fish, amphibians, reptiles, and mammals, there ought to have been countless "transitional stages." Isn't it odd, he asks, that no such transitional fossils are ever to be found, let alone vast numbers of them?

'He adds that, "For a fish to become a land creature, turning its fins into legs would require *new* bones, *new* muscles, *new* nerves—and while it was adapting to life on land, **a** *new breathing system!* Since this supposedly happened from *chance mutations,* which are *rare events,* innumerable creatures would have to live and die during the intermediate period." But, he says: "WHERE IS THE EVIDENCE for these TRANSITIONALS?" He writes,

> "Not in the living world. Among bacteria, invertebrates, fish, amphibians, reptiles and mammals, there are many thousands of species, but NO INTERMEDIATE SPECIES between these groups. ... Evolutionists try to explain the missing intermediates by saying "they all became extinct." But, NO **transitional links** are ever to be seen: NEITHER LIVING, NOR IN THE FOSSIL RECORDS.
>
> "While the fossils show [minor] variations *within* species, they **do not evidence change** from one animal group to another as Darwin claimed [if his theory be true]. For example, while billions of invertebrate fossils exist, fossils illustrating their alleged evolution from simple ancestors are missing." ⁹⁰

'Many liked Darwin's theory, Ali, because it was simple and it was logical. But, *living things are not simple,* nor do simpleness, or logic, by themselves, make something true. The ***pre-Galilean***, theory of **geo-centricity** was logical, but its logicality did not make it true. Similarly, Darwin's theory can account for *some* of the data—but for not nearly enough of it.

'For example, **sexual reproduction**: Darwin claimed that *all* Life flowed from just one (or two) common ancestor microbes, which, having arisen *'by chance,'* from NON-LIVING CHEMICALS!, we are supposed to believe, then, evolved, and branched out, like a tree, into *all* today's species.

'These microbes would, like other microbes, have reproduced ASEXUALLY. But, as John Morrison points out in his brief but excellent book, ***Evolution's Final Days,***

> "Considering the incredibly small probability of having one entity evolve [by pure chance!], what would be the chances of two entities, CO-EVOLVING next to each other, AT THE SAME TIME, and BEING PERFECT for each other? What would be the evolutionary advantage of MOVING AWAY FROM ASEXUALITY [in the microbial world] and needing another member of your species to mate with? [I can't think of any... The idea that such a process could happen by mere chance is even more implausible.]
>
> "The penis and vagina are common in many animals..." but, the "ISSUE IS THE START of it all. **How are we supposed to believe** that some animal in the ocean, which was **asexual**, had a MISHAP IN ITS DNA [random mutation] to [begin to] form a penis, and **another had a mishap to [begin to] form a vagina**? And then the two started having intercourse... And one developed a vagina with ovaries [due to rare, random, genetic typos?] and **another a penis and testicles to start a two sex world.** ... [Due to rare, random, genetic typos?]
>
> "There are a lot of *"how in the heck"* questions here. You will hear a lot of jibber jabber from the evolutionary community but **I want you to *think* with your own mind**, and try to make sense of this logically... it doesn't make sense!" [91] [Yes, we need to think. But, can we be bothered to?! Not sure why, but, often, we don't bother to...]

'I guess, Ollie, a lot of us don't think these things through . . .'

'Well, it can be difficult to challenge the prevailing views. Most of us are not specialists, and if we are told something is true, by the 'experts,' who are we to disagree? Yet, reading Robert Shedinger's ***Darwin's Bluff***, I was reminded, yet again, of the shocking lack of evidence for Darwin's claims: for pure chance to create *Cells or Shells, Eyes or Ears, Blood or Brains, Worms or Moths, Bears or Whales, let alone C O N S C I O U S N E S S, Mind, Emotions, Intelligence, Memory, Curiosity, Love, Joy, Creativity.*

'Darwin wrote: "If it could be demonstrated that any complex organ existed, which COULD NOT POSSIBLY have been formed by numerous, successive, slight modifications, MY THEORY WOULD ABSOLUTELY BREAK DOWN. But I can find out no such case." He could find no such case because, as Shedinger says, he hated to *admit* that his theory had failed. He could "find no such case," either, because did not look very hard, or, because he did not look *at all!* Because there are thousands of such cases!

'Sadly, to this day, his followers behave in much the same way. I found Gunter Bechly's review of ***Darwin's Bluff*** interesting. Bechly describes himself as a one time "card-carrying Darwinist, serving as a fossil curator in one of Germany's natural history museums."

'While there, in 2009, he had the job of mounting an exhibition expressly intended to show that Darwin's work outweighed the works of his modern critics. He writes,

> "To prepare for hard questions from reporters, I decided to give the [Darwin critical] **naysayer books** a quick read, books I had been assured were all froth and foolishness. I soon discovered that **I had been misled**. The arguments in those pages were neither shallow nor illogical. Instead, I came to see that it was **actually modern DARWINISM THAT RESTED ON A CAREFULLY CONSTRUCTED BLUFF.** ... What emerges from Shedinger's deep dive into Darwin's private writings is a picture of a man [rightly] **wracked by doubts and insecurities about his evolutionary theory**, but also a man not above a good *bluff,* one he sold so artfully [to the Victorian public] that he may even have persuaded himself."[92]

'Despite my teenage conversion to Darwin's ideas, I, too, eventually, worked out that his theory—while it was simple, and it could account for *some* of the data, similarities in vertebrate limbs, for example—overall, it DID NOT ADD UP. Darwin and his followers endless attempts to make Life's history conform to his ideas were not supported. His theory was, in the end, a steam-age *"Just So"* story, not a whole lot better than trying to treat the creation story in the bible as a literal description of Life's genesis.

'***Darwin's Bluff*** conveys just how much of a *bluff* Darwin had perpetrated—not just on us, the general public—but on his *scientific* peers, whose professional opinion he so valued.'

'Why?'

'Because he repeatedly presented **On the Origin of Species** to them as merely an ***introduction*** to his theory of 'natural selection,' which, he said, he would, in due course, ***substantiate*** in a much larger, later book.

'What he published in 1859, was he said, just some "BRIEF EXTRACTS" from his larger manuscripts, "**imperfect**," *without* "**references**," only "**general conclusions**," with just a "FEW FACTS in illustration," adding that, no one could feel "more more sensible than [he] of the **necessity** of hereafter PUBLISHING IN DETAIL ALL **all the** FACTS, WITH REFERENCES, on which my conclusions have been grounded; and I hope in a future work to do this."

'Except, his promised "future work" to back up his allegedly very "BRIEF EXTRACTS," "**in detail**," with "ALL THE FACTS,"[93] never came.

'This, though, should not surprise us, because, like his critics, friendly and unfriendly, he realized that his theory, which, initially, he was so passionate about, ***could not explain*** the data of life's history nearly as well, or at all, as he first hoped.

'The transitional (fossil) forms needed by his theory, *are* never to be found—let alone the vast numbers he called for—despite a huge amount of searching since Darwin's time.

'What is sad, is that Darwin's *failed, anti-spiritual, anti-intelligent-design* ideas continue, today, to limit, not just our scientific understandings, but our own, deeper understandings of ourselves and of reality as a whole.'

We continued on our way by the lovely river.

9 Cosmic Soup

'According to science today, all Life needed to start and to evolve were just some **CHEMICALS** + lots of **LUCK**. So, let's look, one last time, at Darwin's 'warm little pond' idea, relating to the very start of Life. Well, many of us may be surprised to learn that there is **no evidence** for any kind of pre-life chemical-soup pond or sea. Biologist, Michael Denton, writes,

> "Rocks of great antiquity have been examined over the past two decades and in none of them has any trace of abiotically produced organic compounds been found . . . **Considering** the way the pre-biotic [pre-life] soup is referred to in so many discussions of the origin of life as an already established reality, it comes as **something of a shock** to realize that **there is absolutely *no* positive evidence for its existence.**" [94] [No evidence.]

'But, even if there *was* a chemical-soup sea, aren't the kinds of chemical reactions which produced some random amino acids in the Miller and Urey experiments reversible? So the forces giving rise to such building blocks would **destroy** them as rapidly as they were formed.'

'Yes, that's true. Miller and Urey needed to solve this problem. How? By intelligently designing a chemical trap to save their building blocks.

'But such a trap would not exist in the 'mindless' or 'dumb' nature of materialist belief, would it?' Queried Ali.

'Correct. Eastman and Missler add that, "The destructive effect of oxygen, ultraviolet radiation from the sun ... makes it unlikely that significant quantities of viable nucleotides and amino acids could ever accumulate in the primitive ocean."

'However, *"even if* they did accumulate... the next step is to explain how they combined to form the self-duplicating DNA molecule and the **thousands of proteins** found in the **simplest** living cells. For the materialist scenario to be taken seriously, it must provide a plausible explanation for the origin of these enormous molecules *without* the introduction *[by spiritual intelligence] of biochemical knowhow*…" They continue,

> 'One of the most difficult problems for the materialistic scenario on the origin of life is something called **molecular chirality**. The building blocks of DNA and proteins are molecules which can exist in both right and left-handed mirror-image forms. This [left and right] "handedness" is called "chirality." . . .
>
> 'In all living systems the building blocks of the DNA and RNA exist exclusively in the *right-handed* form, while the amino acids in virtually all proteins in living systems, with very rare exceptions, occur only in the *left-handed* form.
>
> 'The **dilemma** for materialists is that all "spark and soup-like" experiments produce a mixture of 50% left (levo) and 50% right-handed (dextro) products. ... Unfortunately, such mixtures are **completely useless** for the spontaneous generation of life.
>
> 'For 80 years chemists have been trying to synthesize optically pure mixtures of amino acids in the lab using stochastic **[random]** chemical processes. However, this has never been accomplished. According to physical chemists, it is IMPOSSIBLE…
>
> '**Miller and Urey** acknowledged that the chemical makeup of their experiment consisted of *equal* portions of left-handed and right-handed amino acids.
>
> 'Complex molecules such as DNA and proteins are built by adding one building block at a time onto an ever-growing chain. In a "primordial soup" made up of equal proportions of right and left-handed building blocks, there is an equal probability at each step of adding either a right or left-handed building block.[95]
>
> 'Consequently, it is a MATHEMATICAL ABSURDITY to propose that *only* right-handed nucleotides would be added time after time without a single left-handed one being added to a growing DNA molecule. **Sooner or later an incorrect, left-handed nucleotide will be added. The same goes for proteins**…

Consequently, if even one nucleotide or amino acid with the incorrect "handedness" is inserted into a DNA or protein molecule, the three-dimensional structure will be annihilated and it will cease to function normally.'[96]

'So, given that Life *by-chance* is a MATHEMATICAL ABSURDITY, how come anyone still believes it?'

'Firstly, because we're not told. Secondly, because Life's *deeper* causes cannot be seen. So, we either take the view that: "There cannot *be* any deeper causes," or we allow that: "Logically, they must exist, but they can only be *inferred,* or known, to some extent, by *extra-sensory* means.[97]"

'But our current science does not accept this does it?'

'No. And as to *spiritual* experiences, they are all rejected by this anti-spiritual mindset. People are *deluded* about their own experiences,' tends to be the dismissive attitude. George Wald, a brilliant biochemist, wrote that : "the SPONTANEOUS GENERATION of a living organism is IMPOSSIBLE. Yet we are here as a result, I believe, of spontaneous generation."[98]

'But that's self-contradictory…'

'Yes, but, he continued, "we choose to believe the I M P O S S I B L E," ***not,*** on scientific grounds, but on "philosophical grounds." [99] Remember, Ali, in the science of origins you don't have to be *'right.'* The science of origins, in its current form, is much more *philosophy* than it is 'proven' science. It doesn't have to be *'right.'* No one is going to die if we are wrong.

'We can, like Darwin, make up scientific-sounding *'Just so'* stories, tell ourselves that we have found a *rational*—by which we mean non *super* natural—explanation for Life's origins, insisting that, "No *Intelligent Causes* are needed to explain anything in Nature, not even *Life or Mind or Feelings."* It's not as if we are wrong, a plane is going to crash, or a new drug is going to harm somebody. We can, also, do it quite sincerely. Not because we *know* we are right, as such, but, because **we are so convinced** that that our chosen philosophy is true. So, Life *"has to be"* a cosmic fluke.

'This makes it hard for us to change our views. If anyone refers to the possibility of *intelligent* causation, we immediately conjure, in our minds, a mental picture of a ridiculous, straw man, Father-Christmas-type god-with-a-beard creating and maintaining everything. An absurd representation of *spiritual* intelligence that no one needs to believe in and few do believe in.

'Added to this, many of our materialist friends are angry at *Spirit* because Life can be so hard, at times, and so difficult to understand. Nature, also, can be very harsh, and they feel hostile to the religions as well.'

'Why towards the religions?'

'Because, regardless of their founders original intentions, they have, often, been responsible for so much conflict and suffering. Terrible religious wars were fought, and religious officials used their authority to oppress and to prevent the spread of better ideas. Sadly, this can still happen, even today.

'Others, like Vicki, an old friend of mine, think that for an explanation to be *'scientific,'* it must be materialistic—it must exclude the *super* natural.

'She forgets, like so many others today, that science is about finding out what is *true,* it is not a rigid worldview, like materialist-atheism.

'Hmm...' said Alisha, thoughtfully. 'It's quite an impasse then.'

'Yes, because, right now, there is a conflict between those, like Vicki, and their heroes, like Darwin, who believe that all talk of the *spiritual* is wishful thinking, and those who argue that Darwin's explanations of Life's history have failed, and it's time to let them go.

'For example, modern intelligent design advocate, V. J. Torley, describes how, Professor James M. Tour, "**one of the ten most cited chemists in the world**... [who] has authored or co-authored 489 scientific publications and his name is on 36 patents," took the courageous step, along with over 700 other scientists, "back in 2001, of signing the Discovery Institute's *'A Scientific Dissent from Darwinism."*

'It reads: "We are skeptical of claims for the ability of **random mutation** and **natural selection** to account for the **complexity** of life. Careful examination of the evidence for Darwinian theory should be encouraged."

'A reasonable request, surely. At his website Professor Tour writes:

> 'Although most scientists leave few stones unturned in their quest to discern mechanisms before wholeheartedly accepting them, when it comes to the **often gross extrapolations between observations and conclusions on macroevolution**, scientists, it seems

to me, permit unhealthy leeway. When hearing such extrapolations in the academy, when will we cry out, "THE EMPEROR HAS NO CLOTHES!"?

'. . . I simply **do not understand, chemically,** how macroevolution could have happened. Hence, **am I not free to join the ranks of the skeptical** and to sign such a statement without reprisals from those that disagree with me? Furthermore, when I, a non-conformist, ask proponents for clarification, they get **flustered in public and confessional in private** wherein they sheepishly confess that **they really don't understand either.** Well, that is all I am saying: I do not understand. But I am saying it publicly as opposed to privately.

'**Does anyone understand** the chemical details behind macroevolution? If so, I would like to sit with that person and be taught, so I invite them to meet with me. Lunch will be my treat.

'Until then, **I will maintain that no chemist understands, hence we are collectively bewildered.** And I have not even addressed origin of first life issues. For me, that is even more scientifically mysterious than evolution. Darwin never [formally] addressed origin of life [if we leave off his informal note to Hooker], and I can see why he did not; he was far too smart for that.

'Present day scientists that expose their thoughts on this become ever so timid when they talk with me privately. **I simply can not understand the source of their confidence** when addressing their positions publicly.'[100]

'Torley describes how, in a talk in 2012, Professor Tour said that *no scientist* that he has spoken to understands evolution – and that includes Nobel Prize winners. Here's what he said when a student in the audience asked him about evolution:

'. . . **I don't understand evolution,** and I will confess that to you. Is that OK, for me to say, "I don't understand this"? Is that all right? … Let me tell you what goes on in the back rooms of science – with National Academy members, with Nobel Prize winners. I have sat with them, and when I get them alone, not in public – because it's a scary thing, if you say what I just said – I say, "**Do you understand all of this, where all of this came from, and how this happens**?" Every time that I have sat with people who are synthetic chemists, who understand this, they go "Uh-uh. Nope." . . .

'I was once brought in by the Dean of the Department, many years ago, and he was a chemist. He was kind of concerned about some things. I said, "Let me ask you something. You're a chemist. Do you understand this? **How do you get DNA without a cell membrane?** [protecting the DNA inside the membrane.] **And how do you get a cell membrane without a DNA?** [Which CODEs for the (very) membrane needed to protect the DNA inside.]

'And how does all this come together from this piece of jelly?' We have no idea, we have no idea. I said, "Isn't it interesting that you, the Dean of science, and I, the chemistry professor, can talk about this quietly in your office, but we can't go out there and talk about this?" [101]

'You see, Ali, there is a lot more at stake, here, than practical, everyday *science*. Emotions and convictions are at stake. Reputations are at stake. Yet, as Tour alludes, I am quite sure that many scientists understand perfectly well that Life could **neither start** by chance, **nor evolve** by such means. But, in the current climate, they do not feel safe to admit this publicly.'

'Why not? I thought scientists were always passionate about *truth.*'

'Well, firstly, because they are trained to avoid referring to the *super* natural in their explanations—based on the assumption that the *super* natural is, either, (a) not real, or, (b) if it *is* real, it is not needed to explain anything in 'natural' reality. This is the approach in modern science known as method.o.-logical materialism (MM) or methodological naturalism (MN).

'Secondly, they do not want to put their reputations at risk, even if, by doing so, they are simply pointing out (what should be) the obvious.

'Thirdly, if there is no *Creative Spirit(s) or Intelligent Cosmic Cause(s)* of any kind, guess *who* get to be the new 'gods' of our understandings?

'Not the priests of old, but the natural scientists?' Wondered Alisha.

'Yes. So, many of them, *bluff,* and maintain, in public at least, that these problems don't exist. This is why, in this climate, it takes a strong soul like a Behe, Meyer, or Tour to put their head above the parapet, and to risk the ridicule of those so committed to the current *(Emperor's New Clothes)* views.'

'You know what, Ali? One of the intriguing things about life on our planet, is that for those of us who wish to argue that there is **no** intelligence, **or** love, **or** purpose, in Nature, we can find reasons to think this way—suffering and death. This, in part, is what drove Darwin's views of nature. How could a 'good' god allow so much harshness in the living worlds?'

'I can understand that, Ollie, yet, on the other hand, we can notice the *intelligence and the immense beauty* in Nature, the magic of a child's smile, the kindness and love around us, can't we? It's our free choice . . . isn't it?'

'Yes. Our materialist friends point to the difficulties and the sufferings of this world,[102] and, due to these, they argue *against Intelligence* and *Soul* in nature. Others, like Tour, and Behe, point to the *incredibly* clever ways in which living things are put together, to the miracles of *Life,* to *Love,* to the human *Spirit,* to human *Creativity,* and to the *goodnesses of Life.*

'When it comes to our scientific inquiries, there is also a dichotomy. We can explore, solely, the OUTER aspects of reality, from ATOMS to STARS and everything in between—while ignoring the *subjective, Soul, and inner dimensions* of things. Or, we can inquire, also, into the more *subtle dimensions of Life and of Being—the Life, Soul and Spirit realms.*

'For a long time now, our sciences have focused *solely* on the MATERIAL and the MECHANICAL, on the OUTER-FORM sides, of nature's holistic 'coin,' its 'TAILS' sides, and not at all on the *deeper, inner* causes of things, what we could refer to as the *'Heads' sides, the subjective, inner, spiritual* sides.

'Thankfully, though, Ollie, we still have at least some of the world's earlier peoples, the religions, and other spiritual traditions, to remind us that reality has *Spiritual and Subjective* elements, *all the way out, and all the way in—to the ineffable, yet (somewhat) knowable Mystery,* to which, traditionally, we gave names like *God, Sat-Chit-Ananda, Brahman, Allah, Wakan Tanka, Buddha-Nature, Living Being, the Tao, the Great Spirit, the Great Mystery, Sacred Existence . . .*' We continued on our way by the living stream.

10 Darwin's Mistake

'Evolution is such a broad word that it is often used to mean little more than connected changes over time, versus, sudden, disconnected emergences, like the Big Bang in cosmology, or the Cambrian Explosion in biology. So, if all Darwin had meant was change over time, to say that Life had evolved would have hardly been controversial. To add to the muddle, this broad word has two almost *opposite* meanings, describing two *very different* kinds of processes, both called evolution—one *random,* the other *intelligent.*'

'Do you mean, like the *random* evolutions of landscapes, versus *intelligence* led evolutions, as we see with our ideas and technologies?'

'Yes, landscapes, evolve raNd0mLy. Our ideas evolve intelligently, vengefulness to forgiveness and the golden rule, as do our technologies, Ford model T to modern Fords. So, then, what about *Living Things? Cells, Blood, Muscles, Teeth, Bones, Worms, Fish, Snails, Bears, Whales, People?* Which of the two makes more sense? Rand0mLy or **intelligently?**

'Materialists say, randomly. Their anti-spiritual philosophy leaves them with no other choice. Interestingly, they don't refer, explicitly, to their way of thinking as '**unintelligent design**' or UD. But as they reject '**intelligent design**,' or ID, what are they left with? '**Unintelligent design**,' which describes Darwin's idea that, "**No intelligent input** is needed to explain Living things."

'Darwin argued for a gradual, entropy-defying, probability-violating, rise from the non-existent to the *functional:* from the simple to the ever-more-complex: from LIFELESS CHEMICALS, in a "warm little pond," to living, motile cells. From the *dis-organized,* to the incredibly *clever,* for no reason at all! Yet, nothing we make, not even a basic mousetrap, as Michael Behe pointed out, "falls together" by pure chance. But the most complicated mechanisms we know of, Living things do so? And this, we are supposed to agree, is *scientific?* If people don't like the religions, or the people who run them, terrifying their adherents with tales of hideous hells which only the most sadistic 'spiritual' forces could conceive, or boringly pious heavens, I can understand. Traditional religious depictions of the *supernatural* may be wide of the mark. But the idea that `LIFELESS` chemicals could, by pure chance, turn into *Living Cells*—then, cells-to-worms, worms-to-fish, fish-to-amphibians, dinosaurs-to-birds, caterpillars-to-butterflies, bears-to-whales, apes-to-humans—is clearly ridiculous. All our experience teaches us that clever mechanisms *always* have clever origins.

'So, Ollie, we either follow the classical view, or, we argue, as Darwin did, that *none* of Life's forms, cells, blood, brains, eyes, hearts, lungs, are *intelligently* put together—no, they only *'seem'* to be. Which means that we cannot refer to our *Hearts* as *'clever,'* or to our *amazing Brains* as cleverly designed, because *cleverness* implies `mind` and `creativity`.

'Yes, but our science today, championed by thinkers like Dawkins, claims that there is no intelligence or purpose behind living things. Never mind that **we, as egos,** Dawkins included, cannot make our *Hearts* pump, turn our food into *Energy or Muscle, Heal* our cuts and scratches, let alone operate just one of the *trillions* of stunningly complex *cells* inside of us . . .'

'Our technologies—all far simpler than any Life form—only evolve as a result of our *ideas*. They *never* arise as the only seemingly clever results of the parts being randomly tossed about, upside down and inside out, while we wait to see what wonderful new gadgets pop out, again and again.'

'So, Ollie, do you think evolution, as Darwin described it, took place at all? Bet it (a) randomly as he argued or (b) intelligently, as others argue? Or, did Life unfold in other ways?

'Well, given the Big Bang of the Cambrian Explosion, given the **sudden**, not gradual, appearances of the species, given the long-term **stability** of the species, I don't think we can argue that evolution, as he described it, took place at all—rAnDOmly *or* **intelligently**. Change over time? Yes. Trilobites, worms, snails, fish, dinosaurs, mammals etc all happened. Change over time is just a basic fact of Life. But those changes do not support Darwin's ideas.

'A far better hypothesis than Darwin's is that, at *every moment, Living Nature was, and it is, in-form-ed* by the *same, mysterious, underlying, inherent,* **24/7**, **intelligence** that in-forms everything, including my **living**

body and yours, my mysterious sentiency and yours, ***our amazing minds,*** our mysterious breath—*pneuma*—our beating hearts, and ever circling blood.

'Look at a spider weaving its sticky web, see the gyring birds, or beautiful butterfly. If we do not notice these as intelligent manifestations of *soul and spirit,* we no longer know how to look. It means we have lost all connection with our *own hearts—and the Heart and Soul of all things.*

'The waters are then further muddied by the conflation of two other forms of evolution: **micro-evolution**, which is not controversial, *versus* Darwin's theory of **macro-evolution**, which *is* controversial, and much contested.'

'What's the difference?'

'Micro-evolution describes small variations *within* species—like finch beak variations. Macro-evolution is Darwin's proposal that all living things have descended from just one, or two, universal ancestor microbes. Only, there's **no good data** for any such process. Remember the *Emperor's New Clothes* story? Those who claimed to be able to see his amazing clothes, were also explaining *how* they were made.

'Similarly for Darwin's theory. Those who continue to insist that it looks magnificent claim that its clothes are made of random genetic mutations.

'Random genetic typos, they say, drove Darwin's imaginary process of single-cell microbes morphing into worms, snails, and fish, and, finally, into you and me. Random genetic mutations spun the *non-existent gold* of Darwin's faulty ideas into the spinning spider and gyring falcon, the salmon hunting bear and the mighty whale! Obvious isn't it? Just look carefully.

'As Darwin didn't know about genes, this version of his *Emperor's New Clothes* theory is called neo-Darwinism. Neo-Darwinism proposes a commensurably unrealistic mechanism to drive Darwin's imaginary process of the descent of all species—bacteria-to-bears—with truly massive changes all along the way. This fantastical process for which there is ***no evidence!***

'It seems, Ollie, that most of *'evolution science'* is really just an atheistic creation story for those who wish to deny the *super* natural. But because it is not the kind of **practical science** that *has* to be right, or some important technical thing won't work, a lot of people are not that bothered.'

'Yes. In this area of science, you can (just about) make up any story you like, then, proclaim: "Don't worry we have it all worked out…you, and your ideas, your loves and dreams, are just the **clever-seeming** product of millions of random genetic copying errors! Don't bother thinking it through, just trust us, we're scientists. Life is just an accident and Darwin proved it."

'Forgive the parody, but that's about the size of it. Anyway, according to neo-Darwinism, random, genetic typos, RGTs, turned *single-cell* microbes into far more complex *multi-cellular* organisms like worms, snails, and fish.

'Then, RGTs turned water-breathing fish into air-breathing amphibians.

'RGT's turned dinosaurs into birds, with *completely* different lungs again. Millions of **rare**, (spot the contradiction), RGTs transformed burly bears, or some other mammal, into echo-locating, sea-diving whales.

'RGTs turned shrews into amazingly fast-flying, upside down hanging, sonar equipped bats. RGTs turned caterpillars into beautiful butterflies.

'Only, Ali, there is NO, GOOD, EMPIRICAL, EVIDENCE! for any of these alleged processes! Doesn't this matter? Darwin had an interesting idea. But was it true? He had noticed that by *careful,* clever selection, thoughtful, *clever* people could bring about some **limited** changes to existing species . . .

'Like wolves to dogs or pigeon breeding?'

'Yes. Some call such **within** species changes, like wolves to dogs, [guided by clever people], or minor finch beak variations in nature, micro-evolution. But Darwin made no allowance for the fact that clever human selectors have only ever been able to change a species up to certain **genetic limits** which they have **never** been able to cross without causing damage.

'There's no evidence that micro-evolution, *artificially* guided or *natural,* has ever led to **macro-evolution**, to a new type of plant or animal. But, THIS was Darwin's CORE idea! The species *had* to vary, **massively**, for his theory to be true. Microbe to worm—**huge change**. Microbe to snail—**huge change**. Microbe to fish—**enormous change**. Fish to amphibian—**massive change**.

'Yet, we—clever, focused, human people—despite *all* our efforts, have **never been able to change any species beyond certain genetic limits** without causing damage, let alone creating a new species. That, we have never managed to do. **No massive changes** when it comes to *our* efforts.

'Darwin and his followers have always **ignored** these key points and made unjustified extrapolations from micro- to macro-evolution. Unjustified in theory—because Darwin had NO DATA whatsoever that such massive changes *might* be possible. Unjustified in practice—because there are NO FOSSIL, or living, **transitional species** ever to be found.

'This, given the **strict limits** to the changes which can be brought about by artificial breeding, is not surprising. It has never been possible to *artificially* **evolve** a species, beyond the limit set by its gene pool, without damaging it. Wolves to Chihuahuas? Yes. Wolves to a new species? No.

'Darwin's idea was that if an existing species, like a Wolf, gave birth to a different version of itself, like a Chihuahua, it might survive and become a new species. The evidence, though, is that what Darwin referred to as natural selection, or NS, tends to *confirm,* not to deviate, the species type.

'NS is ***not*** the great, creative, natural *'law'* Darwin hoped he had discovered. No, NS simply allows the existing members of a species, best adapted to current conditions, to survive and to breed more numerously than the less well adapted ones. But it *does not deviate them* to any great degree.

'Feeble or unviable variations of the original kind, like Chihuahuas, will not survive in the wild, and healthy variations will be even more wolflike than before—faster, and cleverer. Yet, as Michael Flannery writes, in ***'Was Darwin a Scholar or a Pitchman (skilled salesman)?,'***

> 'For Darwin, the fact that man could **breed a fancy pigeon** or an especially fast race horse or a unique dog indicated evolution "in action." But, as [his colleague, Alfred Russell] Wallace pointed out to him, [that] when left in the wild, **these fancy breeds** [like Chihuahuas] **either perish or revert to their original type**.
>
> 'Besides, domestic breeding of animals requires the very thing Darwin sought to avoid — **careful thought and pre-selection**. In effect, it **requires a breeding plan and design**. This is clearly not random and purposeless, wholly natural causes operating to produce speciation. **Darwin never saw that logical flaw in his own theory**, yet it was obvious to naturalist Alfred Wallace, zoologist Pierre Grassé, historian Jacques Barzun, and many others.' [103]

'Darwin was unable or unwilling to accept the implications of some of the most obvious flaws in his own ideas. Yet the fossils **never** supported his claims for such vast and "indefinite departure,"[104] as from microbes to worms, snails, spiders, fish, bears, whales, and, eventually, you and me.

'We also now know that cells use **very sophisticated means** to, "protect themselves against precisely the kinds of accidental genetic change that, according to conventional [Darwinian] theory, are the [very] sources of evolutionary variability."[105]

'In fact, all our attempts to breed or to evolve any species beyond a certain point have *never resulted* in a new species arising, but only in existing ones being **damaged**. Domestic dogs damaged by overbreeding and fruit flies deformed in lab experiments are well known examples.

'It is true that in the wild, when conditions vary, minor changes can take place *within* species, due to the potential for *some* adaptive variability built into their gene pools. If the bark on trees becomes darker, as happened during the industrial revolution, the darker moths, *already in existence,* do better than the lighter ones which get eaten more often than the darker ones.

'But when the pollution ends, and the tree barks become lighter again, the fairer moths, **which never completely disappeared**, begin to predominate again. These kinds of **minor**, backwards and forwards **shifts**, have, falsely, been claimed as evidence for Darwinian macro-evolution.'

'Why falsely?'

'Because these minor changes do not depend on random genetic mutations, but take advantage of the moths' *existing* genes, arising from a gene pool which already includes possibilities for darker and lighter examples to hatch, or finches with larger or smaller beaks to hatch, the successfully surviving proportions of which vary as conditions change.'

'But are not the horse fossils and archaeopteryx evidence for evolution?'

'Well, in *The Naked Emperor: Darwinism Exposed,* Antony Latham argues that the horse fossils point to micro-evolution, not to macro. Latham,

> "We do not see, in the horse series, any major change in anatomy, just loss of digits, increased size and altered tooth shape. This hardly constitutes evolution on the grand [macro] scale." [106]

'Jonathan Wells observes that Henry Gee of *Nature* magazine, while himself a believer in Darwin's theory, "candidly admits that we can't infer descent with modification [for horses or any other species] from fossils." [107]

"No fossil is buried with its birth certificate," he wrote in 1999. ... To take a line of fossils [like equines] and claim that they represent a lineage is **not a scientific hypothesis that can be tested** but an assertion that carries the same validity as a bedtime story – amusing, perhaps even instructive, **but not scientific**."[108]

'The primary issue, though, is *not* succession but *cause*. We create, and evolve, things in *successions* all the time. Modern Fords are *conceptual* descendants of earlier Fords. But, every Ford, since Model T, is an *intelligently* created successor model in its own right. It did not evolve by chance!

'So, even if Darwin was right, macro-evolution—microbe to snail, microbe to fish, microbe to spider—would need to be an *intelligently* driven process, not random. But, yet again, it is an *Emperor's New Clothes* story.

'There is simply *no evidence* for Darwin's key idea that one species can gradually turn—be it by purely random means as Darwin's *atheistic* followers believe, or intelligently guided as Darwin's more *spiritually* minded supporters believe—into a completely different type.

'As John Davidson notes, in *Natural Creation or Natural Selection*, "**NOWHERE has any fossilized species** been found which is **on the way** to becoming a spider [or fish or bear or whale or bat or butterfly or anything else]. It's spiders or something else!" [109]

'What frustrates me, Ali, is that so few people now seem to care. Doesn't all this matter? Are we to ignore the data, just to make Darwin *'right?'* This is not how science was supposed to be.

'As to Archaeopteryx, in *Icons of Evolution: Science or Myth?*, Jonathan Wells explains,

'. . . there are too many structural differences between Archaeopteryx and modern birds for the latter to be descendants of the former. In 1985 University of Kansas paleontologist Larry

Martin wrote: "**Archaeopteryx is not ancestral** of any group of modern birds." Instead it is "the earliest member of a totally extinct group of birds.'" [110]

'So, if archaeopteryx is not a credible ancestor for modern birds, is she a good candidate to be a descendent of earlier dinosaurs?' Wondered Alisha.

'Good question. In *Icons,* Wells discusses one idea which is that archaeopteryx descended from a group of dinosaurs the only fossil data for which comes in a stratigraphic period *long after* archaeopteryx was extinct.'

'But how could *much later* dinosaurs be ancestors to an *earlier* animal which was *already* long extinct when they were around?'

'I agree, it doesn't make sense. Wells is also skeptical: "The claim that birds are dinosaurs," he says, "strikes most people—including many biologists—as rather strange. … it defies common sense. Birds and dinosaurs may be similar in some respects, but they are also very different."[111]

"Darwin believed that the species could *vary indefinitely.* In *Darwin's House of Cards,* Tom Bethel calls this "DARWIN'S MISTAKE." It is true that small differences arise between the generations, but, **Darwin's mistake** was to assume that these [minor] differences somehow accumulate over the millennia, so that one species eventually transforms itself into another.

"WITHOUT EVIDENCE Darwin's supporters today *still* [believe this]. ... BUT SUCH A TRANSFORMATION HAS NEVER BEEN OBSERVED [either in living history, or in the fossil records]. **No species [fossilized or living] has ever been seen to evolve into another**.' [112]

We continued on our way.

11 Unintelligent Design *vs* Clever Complexity

'In ***Signature in the Cell: DNA and the Evidence for Intelligent Design***, Stephen Meyer explains how scientists are supposed to search for the *best and most likely* explanation for a given phenomenon. For example, is blind chance ***causally adequate*** to create something? Or is intelligence the better, more *logical* explanation?

'In ***Darwin's Black Box***, Michael Behe gives many examples where mere chance is *not* the most causally adequate hypothesis, because the functionalities involved are not merely complex, but *irreducibly* complex.'

"Irreducibly complex'?'

'He means *functionalities* that must arrive **together**, to arrive at all. Mechanisms which could not emerge in a step by step process of gradually accumulating complexity, as Darwin thought might be possible.'

'Give me an example.'

'We have already met some. For example the cell's highly complex outer wall CODED for by the very DNA *within* the cell which the cell's wall protects. Neither can exist without the other—they *must* arise together or not at all.

'Avian lungs: different to those of all other vertebrates. They could not evolve gradually, without causing death, more or less instantly.

'The *bacterial flagellum*: the tiny tail bacteria propel themselves with. It is powered by a nano-scale motor which, highly magnified, looks remarkably like our own electric motors, mysteriously machine-like in fact.

'Professor Behe explains that this intriguing, nano-scale machine could *never* arise due to a series of random, Darwinian-style, PMC processes. Why not? Because, just like the motors we make, *all* the parts would need to be there at the time of assembly, or it would not work at all.

'So, Behe's hypothesis is that only an in.FORM.ing intelligence of some kind, could give rise to the bacterial propeller, not mere chance?'

'Yes. As he says, there would be no adaptive advantage to just one or two of the flagellum's parts being in place, so the parts would *never* be naturally selected. Remember, regular CHEMICAL LAWS, unguided, cannot generate the *very* **specific**—yet irReGUlar—ordering of parts needed. So Behe's thesis is that, just like the motors we make, the flagellar motor is a product of *intelligence,* an intelligence which is part of the basic fabric of reality, the *spiritual* reality of which our *own* intelligences are parts.

'Ah, Ali, don't you find it utterly absurd that we have to argue for what is so obvious? How can any of us be honestly "intellectually satisfied" with the idea that mere chance can create anything remotely clever?, let alone living things/beings? How is this *scientific? "We cannot physically see Life's deeper sources, so pure chance must create all…"* seems to be the only reply.

'I agree, it is very odd…'

'Some of those, Ali, who cannot see what seems obvious to the rest of us, have tried to rebut Behe by arguing: "This or that part of the bacterial motor *might* have been part of something else, and, it *might* have become a part for the flagellum, and this *might* disprove Dr Behe's theory."'

'But that just piles more unlikely speculation onto unlikely speculation.'

'Yes! In *No Free Lunch: Why Specified Complexity Cannot Be Purchased without Intelligence,* William Dembski shows how mere luck is incapable of generating the totally **specific** sequences of parts and software which define all biological mechanisms. [Ah me, why do we still need to argue for these so very obvious points? It feels so absurd.] With an irony, which seems to escape them, some of Darwin's modern supporters, like Richard Dawkins, have **intelligently designed software** to try to show how living mechanisms could, they say, evolve by 'rA*nd*Om,' no-intelligence-needed processes.

'Dembski shows, however, that these computer scenarios all fail. Why? Because they smuggle in pre-defined search criteria which are *neither* truly random, nor *'unintelligent,'* nor available in the 'directionless' and 'purposeless' Nature of materialist belief, while deluding themselves, and their audiences, that they do not. A reviewer of *No Free Lunch* writes,

> 'Dembski describes the 'shell game' [3 cups trick] whereby **evolutionary programmers try to obscure the 'smuggled in'**

information and claim that their programs have generated the output[ted] information from scratch. ...

'My only hope is that critics will have the **decency** to actually engage the ideas presented here. ... Dembski has ... 'taken on' the established Darwinian orthodoxy, demonstrating **the insufficiency of non-intelligent processes to generate the specified [non-random] complexity** [that defines all Living things].' [113]

'As Dembski says, neither regular chemical laws, unguided, nor mere chance, can generate the **totally specific** sequences of DNA and parts which make possible all the biological mechanisms which Behe discusses.

'In *Billions of Missing Links: A Rational Look at the Mysteries Evolution Can't Explain*, G. S. Simmons gives many more examples of biological complexity which *cannot be explained away* by hand-waving references to *'evolution,'* or to *'lucky'* mutations, or *'natural selection.'* For example, the astonishingly clever defense technologies of some insects:

'An Australian termite called the nasui ... If provoked by a predator, it will swing its head side to side and squirt loops of a chemical mix. One component **will glue** the attacker in place, another will **paralyze it,** another may kill it, and a fourth will attract other termite soldiers to gather around the intruder and to keep spraying it. Their sprays contain alpha-pinene, beta-pinene, limonene, trinervitrenes, and kempanes. The latter two chemicals cause the stickiness and they have **not been found** anywhere else in nature. This species also **defies [Darwinian] evolutionary logic**.' [114]

'The bombardier beetle's weaponry is just as extraordinary.

'This African insect can fire off two chemicals, hydrogen peroxide and hydro quinone, **from separate storage tanks and rear jets**. [The beetle can fire accurately and repeatedly. While separate the compounds are harmless. **When mixed they instantly heat up and hit their opponent at boiling point**, 100°C!] When the chemicals combine, they form a new chemical that burns the predator. The beetle can shoot these chemicals with an uncanny accuracy, as well, to either side, backward, or even forward, by swinging its tail under its abdomen. **Special nozzles blast predators at a rate of 500 bursts per second**, each at a speed of 65 ft./s. These chemicals are potent enough to severely damage a mouse and injure the eyes of any animal . . . Yet these chemicals are entirely benign when stored separately at the back end of these beetles. **How could this happen by accident?** "Oops, those two chemicals didn't work" (spoken by an intermediate species).

"Mind if I try two others before you eat me?" Or, "Could you stand a little taller so I can get you with my nozzles?"

'Keep in mind there are hundreds of thousands of chemicals on this planet to choose from. And even if the combo turned out perfectly right the first time, the beetles still needed a way to make them, store them, and fire them off. [115]

'Are these all **merely lucky** chemical mutations that protected the species? ... How did the **special storage glands** that protect the owner evolve? Before or after? All at the same time?

'What happened to the thousands of **missing links?** [There is no fossil or other evidence whatsoever for any such links.]' [116]

'Simmons says of the Bombardier's defense system,

> 'The entire mechanism defies [Darwinian] evolution. **How** and **why** would there have been a gradual development of chemicals that were **useless** until they finally reached their present form? This system is **all or nothing**.' It is irreducibly complex.

'Michael Behe adds to these criticisms of Darwinian evolution,

> 'There is **no publication** in the scientific literature in prestigious journals, specialty journals, or books that describes **how** molecular evolution of any real, complex, biochemical system either **did** occur or even **might** have occurred. There are assertions that such evolution occurred, but absolutely **none are supported** by pertinent experiments or calculations.' [117]

'It reminds me, Ollie, of Professor Tour's comment, "I simply **do not understand, chemically,** how macro-evolution could have happened..." without a *guiding or informing intelligence* of some kind being involved.'

'I totally agree. Here's another interesting thing, Ali. Did you know that what we think of as *Darwin's* theory, actually, had *two* inventors, and it was, initially, thought of as belonging to both of them, not just to Darwin.

'No . . . who was the other one?' Alisha queried, looking intrigued.

'Alfred Wallace. Initially, Wallace's ideas were very similar to Darwin's. But he, unlike Darwin, eventually realized that **no pure chance based theory** of living things, including Darwin's, will ever work.

'Interestingly, Wallace also carefully investigated *spiritualism* and afterlife communications, and, along with other 19th century luminaries, like Sir Arthur Conan Doyle (literary genius, *Sherlock Holmes*), Sir Oliver Lodge F.R.S. (world class scientist, spark plugs), Professor James Mapes (chemist), Frederick Myers (classicist and dedicated psychic researcher), Sir William Barrett F.R.S (physics), Professor Camille Flammarion (founder of the French Astronomical Society), he concluded,

> 'that the phenomena of *Spiritualism* in their entirety do not require further confirmation. They are proved quite as well as facts are proved in other sciences.' [118]

'Apart from the modern *psychic and spiritualist* research data which falsified materialism, there were many other things, Wallace realized, which Darwin's theory did not explain.

'Like. . .'

'Like the magical metamorphosis of the crawling caterpillar into the beautiful Butterfly.[119] Wallace worked out—again it didn't take a genius—that there was *no way* mere chance could account for the extremely complex transformations involved. A caterpillar is a viable organism in its own right.

'There's **no reason** for the [rAnDom] caterpillar of Darwinian thinking to *'randomly'* dissolve itself into a life-forceless[120] and *'purposeless'* chemical-fluke soup, then, *'accidentally,'* re-emerge as a beautiful Butterfly. It is utterly *unrealistic* to claim that some, random, genetic typo could cause a caterpillar to mutate, *just a little,* towards making a cocoon. Even if a caterpillar did, by (impossible) chance, start to make a cocoon, what would happen? Let's think it through to avoid accusations of intellectual laziness.'

'Hold on, Ollie. Living cells, we now realize, protect against random genetic errors, by very clever means. Such a dramatic change, as even *starting* to form a cocoon, would require **millions of mutations**, not just one or two. Then, even *if* such a wildly unlikely thing happened, *why* would it be naturally selected? It would confer no survival advantage, let alone lead to a beautiful new Butterfly, but only to ***death!***

'There'd be nothing to cause the ex-caterpillar to return to its prior caterpillar condition, let alone forward to a never-before-seen butterfly. Even if this happened, more than once, for any caterpillar starting to mutate, **only death would await.**' Ali frowned. None of it made sense. Why was so much nonsense put forward as scientific in the area of origins?'

'I agree. There is no survival advantage in merely starting to form a cocoon. What, after all, would cause the, 'it-used-to-be-a-caterpillar,' now, 'it's-a-useless-soup-in-a-cocoon,' to morph into a beautiful Butterfly, necessitating *millions* of complex biochemical changes, not just one or two?

'No blind process could cause the *'purposeless soup'*[121] the Caterpillar had dissolved into, to turn into a Butterfly. It is *absurd* to suggest that **random genetic typos** could give rise to the miraculous bio.logical happening that is a gorgeous Caterpillar or Butterfly. Bio = *Life!*

'Logos = *the Word!*, the *Soul* of in-**FORM**-ation, to come into form. Darwin had observed that skillful breeders could modify existing species— like tulips, pigeons, or dogs—to a **strictly limited** degree. But to extrapolate from this, as he did, that one species could change into a wholly other kind —via common descent, with huge changes all along the way—due to *rare, genetic typos!*, was unjustified by the data, then or now.

'Darwin, not having our knowledge, did not realize that Life could *never* arise, or evolve, by chance. Had he known this, he would, surely, have conceded,

> 'I realize, now, that the idea I mentioned to my friend Hooker, that Life might have arisen, by chance, in some warm little pond, is **impossible**. It has been repeatedly shown that such transformations

as the caterpillar-to-butterfly process, *could not possibly* 'have been formed by numerous, successive, slight modifications.'

'This means my theory absolutely **breaks down**. To try to explain all of Life as nothing more than a very long series of extremely *unlikely* chemical coincidences no longer makes sense.

'In light of what we now know, we must admit that there is a *tremendous* intelligence, to the 24/7 functioning of Living things.

'It would be *unscientific* to deny it. However mysterious this may seem, **intelligence** and intelligent causation ***must be*** parts of the basic fabric of reality. The idea that *Being-Existence, that **what-holistically-Is**,* is MINDLESS to its core, and that *bright, intelligent minds,* like yours and mine, could arise by MINDLESS coincidence, no longer makes sense—if, indeed, it ever made sense.

'This is why, if we are to make any further progress on *Life's* origins, in the intellectual west, we must accept this, even if it feels like a blow to the pride of our prior beliefs—our onetime *faith* in blind chance as a supreme creative principle.'

'Darwin wrote, in *The Origin of Species,*

'If it could be demonstrated that any complex organ existed, [*Cell, Blood, Brain, Tongue, Eye, Heart*] which ***could not possibly*** have been formed by numerous, successive, slight modifications, my theory would ***absolutely break down***. But I can find out no such case.' [Oh, dear Darwin, it is not hard to find thousands, if you look!]

'In the light of our modern knowledge, Ali, it would make as much sense for scientists today to confidently proclaim that,

'If it could be demonstrated that any complex, AI guided Robot, Car, TV or Computer, [all far simpler than any Living thing], **could not possibly have been formed** by numerous, successive, slight modifications, driven by nothing more than the **laws of random swirling and mixing**, in combination with numberless lucky *natural* coincidences, our theory of evolution by unintelligent design or UD **would absolutely break down**. But we can find out no such case!'

'Such a statement would be absurd. But, when it comes to living things, the usual rules no longer seem to apply. It is quite clear, by now, that there is nothing in the laws of chemistry—unguided—to cause living things to arise.

'In 1859, Darwin put forward a dramatic new hypothesis. All Life had descended from just one, a-sexual, universal, ancestor, microbe: **descent with huge** modifications, *if*, such a microbe **could vary** more or less **infinitely…**

'A theory of life which the late philosopher, Daniel Dennett referred to in the title of one of his books as *'Darwin's Dangerous Idea.'*

'Why *'dangerous?"*

'Because Darwin had questioned so many previous cultural certainties. Given the physical invisibility of Life's deeper sources—rather than continuing to acknowledge all the blazingly obvious evidence for intelligence, purposefulness, and design in nature, and take the classical, *spiritualist* view—he wondered whether Life *had* any deeper sources?

'Maybe there was no *Intelligent Cause (or Causes) of Anything?*

'Maybe, long ago, there was a 'warm little pond,' with some CHEMICALS in it, some flashes of lightning, and, *Voila, Life!* Darwin did not know enough biochemistry to realize that this was a chemical impossibility and, as Eastman and Missler said, a MATHEMATICAL ABSURDITY.

'Yet, to try to make his own views on origins seem true, Darwin **had** to *disregard* all our real-world *experience,* scientific and everyday, which tells us that no complex mechanism *ever 'falls together'* by probability-defying, entropy-negating chance—by **'unintelligent design.'**

'Perhaps the real danger in Darwin's idea, unnoticed by Dennett, and so many other academics and intellectuals, in the years since Darwin, was in the **harm** it would eventually do to science and to rationality?' Reflected Alisha.

'I agree, Ali. The pretense that there is no intelligence or purpose to how living things are made, or function, *24/7,* is so clearly *untrue*. It is, also, so clearly not true that mere 'luck' can create anything remotely functional.

'Unfortunately, after Darwin, the denial of *intelligence* and *design* in Living things, and of the *super* natural in general, in the name of rational, logical science, *did make* our science culture, in the area of origins, at least, totally *irrational*. It led to a breakdown in *spiritual* belief, and left us with a science adept at measuring and harnessing many things, but hopeless at seeing the *true* values and the *inner* qualities of anything.

'It is very clear by now—even allowing for the physical invisibility of Life's *deeper* sources, which, most of us, can only *infer,* not see directly, unless we have *spiritual* vision—that so called MATERIALISM does not make the best sense of the data overall.

'Yet, for this **dreary, inquiry-restricting, science-distorting, reality-shrinking philosophy** to continue to seem *true,* so much logic, intuition, and, for those who have them, *psychic, spiritual,* and *mystical* experiences, have to be dismissed, ridiculed, or denied, because, if this is not done, 'm..a..t..e..r..i..a..l..i..s..m' *is falsified.*

'We, though, who know so much more than Darwin and his contemporary supporters, can no longer justify thinking as they did. If, despite what we now know, we continue to do so, it must be for reasons of ideology, not ones of science.

'I guess, Ollie, it's partly because many today don't like the religions and their officials. Or, because they are angry at their idea of god? Or, because they'd rather we pretend that we know what Life is, and how it begins, when, scientifically, we still don't have a clue, because, as we said before, explaining Aristotle's first two causes, '**made of**' and '**how it works**' does not tell us how or why something arises in the first place.'

'I agree. Ironically, *spiritually* we already have more understanding. Namely, all our *spiritual* knowing tells us that intelligent Life has intelligent causes. Why is that such a controversial statement? In our strange times, it seems to be. Anyway, the key point is that we have, today, not just one, or two, but *many* demonstrations of the kind which show that Darwin's materialist-atheist theory of Life's origins has "absolutely broken down."

'Here are just a few. There are many more. With just a little reflection, we can, all of us, find plenty more examples of our own.'

1. **Life,**[122] **and Living Cells,** the mind-boggling complexities of which Darwin and his Victorian contemporaries knew so much less than we.

2. **Irreducible Mind.** *Immaterial, Intelligent, Curious, Creative Mind* cannot be reduced to MINDLESS MATTER.[123] It should be obvious.

3. **The *irreducibly* complex bacterial flagellum** and its nano-scale motor,[124] of which Darwin and his contemporaries knew nothing.

4. **The *irreducibly* complex biochemistries** of vision and of the blood clotting cascade discussed by Behe,[125] of which Darwin knew nothing.

5. **The *irreducibly* complex** defense technologies of insects discussed by Simmons in his book *Billions of Missing Links*.

6. **The** amazing ***metamorphosis*** of the Butterfly considered by Wallace.

7. **Life's *incredibly complex* DNA CODING** which, like all code, could *never* arise by mere luck—and, of which Darwin knew nothing.

8. **Life's *thousands* of specifically ordered** protein sequences, which could never arise by chance—it's not merely unlikely, it's impossible.

9. ***Bird's Feathers* and *Avian Lungs*,** neither of which could be created by random genetic (copying error) mutations (Denton).

10. **The *evidence of the all-important fossils*** which refutes Darwin.

11. **The** data of SUDDEN **species emergence** which refutes Darwin.

12. **The** tremendous ***stability*** of the species which contradicts Darwin.

13. **The** evidence from ***embryology*** which contradicts Darwin.

14. **The** evidence from **genetics** and **molecular biology** which refutes Darwin's tree of life.[126] It's not a tree of life, as he envisaged it, but a wonderfully interconnected ***web*** of Life.

15. **The** evidence of the ***world's most skillful breeder*s** that it has never been possible to change any species beyond certain ***genetic limits*** without causing damage or death—never mind a new species.

We continued on our way.

12 Is Sudden Appearance Evolution?

"The ABSENCE OF FOSSIL EVIDENCE for intermediary stages between major transitions in organic design, indeed **our inability**, even in our imagination, to construct functional intermediates in many cases, has been a PERSISTENT ... PROBLEM for gradualistic [Darwinian] accounts of evolution." So wrote world renowned fossil expert, S. J. Gould[127]

"DARWIN'S PREDICTION of rampant, albeit gradual, change affecting all lineages throughout time is REFUTED. The record ... speaks for **tremendous anatomical conservatism**. Change in the manner Darwin expected is **just not found** in the fossil record." World class fossil expert, Niles Eldredge.[128]

"[It] remains true, as every paleontologist knows, that most NEW SPECIES, genera, and families, and that nearly all new categories above the level of families, APPEAR in the [fossil] record SUDDENLY . . ." World level fossil expert, George Gaylord Simpson.[129]

'These statements, Ali, by world class fossil experts, all *falsify* Darwin.'
'Should we not listen to them then?'
'Of course. But, don't be fooled, they all remained Darwinian in spirit.'
'Why? The fossils, they said, did not support Darwin.'
'Yes, but, to their way of thinking, their statements didn't change very much. They still wanted to believe in Darwin overall, and they continued to look for wholly materialistic theories to try to explain the data. Gunter Bechly quotes from *The Comprehensive Guide to Science and Faith*:

> "Every theory makes certain predictions." Darwin's core prediction was **"gradualism,"** that all the changes in the history of life were due to **"a continuous accumulation** of small changes" over vast time. "Because Darwin wanted a **wholly naturalistic** explanation" for life's history, whereas, sudden changes in organisms **"would**

require miraculous events. Therefore, he mentioned not fewer than **six times** in his magnum opus *On the Origin of Species* the Latin phrase *Natura non facit saltus." 'Nature does not make L E A Ps.'*

"This claim is still made by Darwinians today. The most well-known modern popularizer of Darwinism . . . Richard Dawkins, wrote in his 2009 bestselling book, *The Greatest Show on Earth* . . :

"Evolution not only is a gradual process as a matter of fact; it **has to be gradual** if it is to do any explanatory work." [my emphasis]

"GRADUALISM is NOT an "optional element of Darwinism," but it is essential for its success as a materialist explanation for the origins and diversity of life. If GRADUALISM is WRONG Darwinism is falsified."[130]

'However, rather than accept that their discoveries had falsified Darwin, Eldredge and Gould tried to *rescue Darwin* from the fossil facts, with an idea which they called punctuated equilibrium.

'PE tried to account for the fact that, by contrast with Darwin's claims, Life is **stable** for **long periods**, followed, at times, by **rapid change**, and by the mysteriously **sudden,** not gradual, appearances of new species.

'However, the problem with PE, is, that the limited numbers—and generally harmful effects—of random genetic mutations **could not possibly lead** to 'descent with modification,' to the **vast changes** in form and function needed to make either PE or Darwin's far more gradualist ideas work.

'Random genetic mutations, as Stephen Meyer explained in ***Darwin's Doubt***, just do not have the necessary creative power. For Darwin, evolution needed **"infinitely numerous transitional links"** forming **"the finest graduated steps."** He said "nature does not make leaps."[131] The fossils show the **opposite!** New species appear in a very **abrupt**, non-evolutionary, way.

'To those who say that earlier forms gave rise to diverging families of species, there is NO EVIDENCE for such hypothetical ancestors. As Colin Patterson explained to Tom Bethel, the notional spaces on the notional branches of Darwin's notional tree of evolution are *non-notionally* empty.

'There are imaginative drawings of Darwin's imaginary transitional species, and the *faith* that Darwin's shared ancestors 'must' have been there, because, without them, his theory *fails*. But this puts the cart before the horse, and, we can't just ignore the fact of sudden appearance.'

'That is odd, though... how could the species 'suddenly' appear?'

'Well, consider the Big Bang. An E N T I R E   U N I V E R S E appearing from no-where to now-here. It's not some slow evolutionary thing.'

'So some things just suddenly emerge...'

'That's the data: sudden appearance—no plausible shared ancestors—no transitional species (let alone vast numbers of them)—the strict limits to genetic change—which the most skillful breeders have never been able to cross—the data from embryology and genetics. All conflict with Darwin.

'What do you propose, then?' Queried Alisha, looking thoughtful.

'I propose we work with **the data we have**, not the data **we'd like to have**, whether we are supporters of Darwin or young-earth creationists who try to impose Bishop Ussher's 17th century ideas onto the evidence.

'The evidence is that, in conflict with Bishop Ussher's early modern ideas, Life on Earth is far older than the few thousand years he allowed for, based on his methodical but far too literalistic readings of the bible.

'The evidence is that, in conflict with Darwin, the **species *did* suddenly** emerge, from seemingly no-where to now-here, in a mysterious pattern of successive Life Waves, followed, at times, by mass extinctions. While this doesn't give us a new theory of Life, it *does* free us to consider the evidence as it *is,* and not try force it to fit either Bishop Ussher or Darwin's ideas.

'It is also useful to remember that in deep reality, as Max Planck discoverer of quantum physics pointed out, there is **NO MATTER** as such. There is just e . n . e . r . g . y   s . p . a . c . e 99.9...9...9...9...9...9...9...9...9...9...9...9% of it, and everything, *this* amazing second and *every* amazing second, is disappearing and reappearing, de-materialzing and re-materializing, out of the quantum ground, now here, now not here, *trillions* of times a second.'

'I'm not sure what you're saying here...'

'That *all* things, Ali, having FORM and imbued with *Living Being,* are emergences from the *Original, Form-less—It-Just-Is—Unseen,* to the **IN-FORM-ED,** to the (only partly) **MATERIAL** and mechanical.

'Some in-FORM-ings, like the Big Bang, occur very rapidly, some, like the Cambrian big bang, less rapidly, and, others take slower, more sequential forms. The key question for us, though, is not in.FORM.ation the data of which are all around us, from **atom to GALAXY** and everything in between, but are the Plants, Animals, and People intelligently in.FORM.ed phenomena, or are they just *Flower–Shaped, Animal–Shaped, and People–Shaped,* coincidences of CHEMISTRY, with no more *Soul* or meaning to them than to the rocks randomly strewn about this beautiful valley?

'Which makes more sense, which is the more logical, overall? Do we live and move in a mysterious but dumb universe, as so many now believe, or in a mysterious, but intelligent one, as powerful thought leaders like Plato, Buddha, Lao Tzu, Jesus, Galileo, Newton, and Planck all held?'

'Are we part of an intelligent cosmic mystery or MINDLESS?' Ali mused.

'Yes, if the *amazing, blazing,* materialization of an entire universe, from no-thing to some-thing, was possible, then other, sudden, seemingly, non-evolutionary emergences may also be possible.

'Our most basic question: do Life's amazing expressions come from an **Intelligent** Universe Generating Mechanism (UGM) which *'just is,'* or from a MINDLESS Universe Generating Mechanism (UGM) which *'just is?'*

'Each is as mysterious and as *'impossible'* as the other, because, be the *Final Source of All intelligent?,* or, be it *mindless?,* **how** come it exists? How **can** it exist? How can there be a *cause* of anything at all, rather than nothing at all? Even one atom? No one knows the answers to these questions. All we can ask is, which of the two types of *Ultimate Causes, intelligent vs* MINDLESS, is, all things considered, the more likely?'

'The answer seems obvious to me, Ollie. Our materialist friends say there cannot *be* any such causes, because life's deeper causes are not visible: "We can't *see Spirits or Souls* in our telescopes!" Our *spiritualist* friends accept that Life's deeper causes are not physically visible, but they **infer** them, using logic. They say, "Open your eyes! Look at a child's smile, or at your lover's eyes. Look at the eagle's wing or the bee's sting, consider the marvels of Blood, Livers, and Lungs. It is absurd to claim that any of these amazing things arose by chance."

'Ah, dear Alisha, I do so agree. Why don't more of us see? Meanwhile, the fossils disagree with Darwin, as they always have. Firstly, because of,

'**Sudden Appearance** – and secondly because of,

'**Stasis** – most species exhibit **no directional change** during their tenure on earth. They appear in the fossil record looking much the same as when they disappear; morphological **change is usually limited and directionless.**'[132] So wrote S. J. Gould.

'In 1859, Darwin published what were, he said, just some "BRIEF EXTRACTS" from his larger manuscripts, "**imperfect,**" *without* "**references,**" only "**general conclusions,**" with just a "FEW FACTS in illustration."

'These 'extracts,' from his promised much larger, later work, can no longer be considered good science—especially as his promised "future work" to back them up, "**in detail**," with "ALL THE FACTS,"[133] never came.

'S. J. Gould pointed out, in 1980, that the updated version of Darwin's theory, called neo-Darwinism, the theory of evolution by random genetic mutations was also *"dead,* despite its persistence as textbook orthodoxy."[134]

'Antony Latham writes, "the general situation found in successive strata is that of *stasis,"* not Darwinian evolution. On the back-boneless animal body plan or anatomical design, called echinoderms, to which Star Fish and Sea Urchins belong he comments,

> "There are NO TRANSITIONAL FORMS found in the fossil record that link the various types of echinoderm since their first appearance in the Cambrian period. What we see is the initial phylum [anatomical design] APPEARING SUDDENLY around 530 million years ago – in a wide variety of forms . . . **I have searched hard** for any evidence of **transitional forms** between the major groups which exist today and I HAVE SEEN NONE. What we do see is **[just] small changes** in existing groups (micro-evolution).

'The initial phylum appears along with all the others in the Cambrian explosion, it is fully complex and has more classes than we have now; the classes that survive appear suddenly also. The only clear evolution is on the micro-scale. It is the **sudden**

appearance of new forms without linking transitionals that remains the real mystery – just as it was in Darwin's day.

'All that is written about the echinoderms is mirrored in the fossil record of the other invertebrates. I have used the echinoderms as one example of what is found also amongst fossils of the other main phyla of invertebrates: arthropods, mollusks, sponges, cnidarians, bryozoans and brachiopods.

'**Check any textbook on the subject** and you will see that SUDDEN APPEARANCE and DISCONTINUITY is the norm. The problem that Darwin so clearly saw has not gone away with time and THE FOSSIL RECORD seems now to REFUTE THE ENTIRE BASIS of his theory of gradual changes." [135]

'In 1979 Luther Sunderland wrote to then senior fossil expert at the Natural History Museum in London, Colin Patterson, to ask him why, in his then recent book, ***Evolution***,[136] **he had *not* included a** SINGLE PHOTO of a transitional fossil demonstrating Darwinian evolution. Patterson replied with this refreshingly unusual and honest admission.

'I fully agree with your comments on the lack of direct illustration of evolutionary transitions in my book. **If I knew of any, fossil or living,** I would certainly have included them. You suggest that an artist should be used to visualize such transformations, but where would he get the information from?

'**I could not, honestly, provide it**, and if I were to leave it to artistic license, would that not **mislead the reader**?'

'Patterson added:

'[S. J.] Gould and the American Museum people are hard to contradict when they say THERE ARE NO TRANSITIONAL FOSSILS. ... You say that I should at least "show a photo of the fossil from which each type of organism was derived." **I will lay it on the line**—there is NOT ONE SUCH FOSSIL for which one could make a watertight argument.'[137]

'"**Not one such fossil**. . ." So, like Eldredge and Gould, this was another world-class expert admitting that the data didn't support Darwin?'

'Yes. Patterson was following the data. This is what Eldredge and Gould were attempting to do with their theory of punctuated equilibrium.'

'But you said they were **trying to *force*** the fossil data to *'fit'* with Darwin's ideas to try to rescue Darwin's theory from the uncooperative fossil facts. What was their idea precisely?'

'That the species will be *stable* for long periods, *without* changing in the way Darwin predicted, but, when evolutionary change *did* occur, *it was so rare* and of *such short* duration that *no fossil evidence* can ever be found for it! Their solution, to the contradictions to Darwin they themselves had found, may sound odd, but Eldredge and Gould were not anti-Darwinian.

'They were just trying to **keep their Darwinian cake** *and* to eat it. They admitted that there are "NO LINKING FOSSILS," which, in any less ideologically contested area of science, would have been enough to *refute* Darwin, and to allow us all to move on. They were, however, reluctant to follow the logic of their own discoveries and to let Darwin go.'

'So why did they continue to argue that Darwin was *'right,'* despite their own findings that he was wrong? Shouldn't they have started over?'

'Well, they claimed that when the species *did* transform—RaNd0m microbe-to-random-worm, worm-to-fish, fish-to-amphibian, dinosaur-to-bird, bear-to-whale, shrew-to-bat, ape-to-man—there were *so few* transitional forms that no fossil examples of them can ever be *found!* Darwinian evolution *did happen, honestly, it did!,* but in such HUGE LEAPS that there were **just too few** linking species for there ever to be any preserved transitional fossils showing Darwin's imaginary process in action!'

'That sounds ridiculous to me, Ollie. Talk about avoiding the obvious…'

'I agree. Also, problematic for Eldredge and Gould, is that they had **no** plausible mechanism to drive the very **rapid** changes they argued for, *versus* Darwin's very slow—nature does not make **jumps**—form of evolution.'

'Isn't the issue, though, that the idea that *Life* is *spiritual* in origin, that it is truly *miraculous,* is unacceptable to science today? So many academics have given up on *spirituality* and they think that Darwin's ideas are the only possibility. They will never consider *super* natural causation…'

'Maybe. Yet, their own, absurdly unrealistic, "pure-chance-can-create-anything" dogmas require us to ignore that *every, humble, cell* in our bodies, *all our senses,* all *our organs,* are vastly cleverer than anything we can make —using all our powers.

'Darwin's materialistic "it's-all-just-dumb-luck" paradigm will never work to explain *Life* and *Mind* on our planet.

'As science writer Richard Milton notes in ***Shattering the Myths of Darwinism: a Rational Criticism of Evolution***,

> "Darwin's theory of evolution relies on the idea of random genetic mutations… altering species... But… the extent of genetic change wrought by natural selection is quite limited…
>
> "[It is also an] embarrassing fact that… NO TRANSITIONAL SPECIES SHOWING EVOLUTION IN PROGRESS BETWEEN TWO STAGES HAS EVER BEEN FOUND… the theory of evolution has become an **act of *faith*** rather than a functioning science…
>
> "Not until the scientific method is [properly] applied to Darwinism will it be exposed, and only then will the right questions be asked about the mystery of life on earth."[138]

We continued on our way.

13 Fabulous Bats

'In *Evolution, a Theory in Crisis*, biologist Michael Denton writes,

> "The difficulty of envisaging how evolutionary gaps were closed [by Darwinian processes] does not stop with birds: take the case of the bats. **The first known bat** which appeared [abruptly] in the fossil record some 60 million years ago **had as completely developed wings as modern forms**." He continues, "As in the case of birds, how could the development of the bats' wings and capacity for powered flight have come about gradually?" [139]

'Bats, Ali, like most other species, appeared abruptly on our planet.'

'No fossils of un-bat-like species on the way to becoming bat-like?'

'No. There are no fossils of wingless rodents gradually turning into fast flying, *sonar* equipped bats, as Darwin predicted, "if his theory be true." There are assertions as to such transformations but *no evidence* for them.

'As Anthony Latham notes in *The Naked Emperor, Darwinism Exposed*,

> '**What is not in any way explained**...[by Darwin are] the *sudden appearances* of completely new forms and structures – indeed, *all* the appearances of the phyla and classes (higher taxa). These remain a **total mystery** and **do not fit in** with any known mechanism or theory.'[140]

'In *Billions of Missing Links: A Rational Look at the Mysteries Evolution Can't Explain*, G.S. Simmons writes,

> 'Biologists speculate that the bat's ancestors might have been tree dwelling shrews or moles who had some capacity to glide, but that's **mere conjecture** [a Darwinian *"Just So"* story]. **It doesn't explain echolocation,** which is their [amazing] **sonar ability** to find food on the wing at night, or how they can purposefully drop their

body temperatures at night to save energy. Most mammals constantly burn energy to maintain a specific temperature, 24/7.

'... compared, by size, to our sonar detectors, theirs is one trillion times more efficient. **Bats can fly at speeds up to 60 mph and at heights of 10,000 feet**. Often they live in caves by the millions. Nurseries may have as many as 2000 pups per square meter, yet mom always seems to know her offspring.

'One interesting capability that breaks the rules of [Darwinian] evolution is the bat's **ability to lock onto a rock and hang upside down without falling**. This is made possible by an automatic locking mechanism in their feet. A tendon closes the toes against the rock, and then the bat's weight creates the proper tension downward to keep it locked. The moment the bat grabs the rock the clamping mechanism snaps in place.

'**The unprecedented backward facing knees with forward facing** feet aid the process. Bats will sleep and carry on most of their necessary activities in a hanging position for hours, only flipping 180° whenever there's a need to evacuate their bladder or defecate. **One might feel sorry for the predecessors** who had not yet learned to flip over.

'When it's time to hunt for food they just they just release, drop, and take off. **Partially-hanging, sometimes-falling intermediates did not exist**.'[141] [There are no transitional bat species.]

'Darwin said that, "if his theory be true", the fossil records ought to show existing species very G R A D U A L L Y turning into new species,

"by **differences not greater than** we see between the natural and domestic varieties of the same species at the present day. [… and that] the number of intermediate and **transitional links**, between all living and extinct species, must have been **inconceivably great**. But assuredly, *if* this theory be true, such have lived upon the earth."[142]

"*If* this theory be true," Ollie. Yet, so many of us ignore that it's not true. Why don't we call out Darwin's *Naked Emperor?*

'Yes, but is it fair to blame the average person? How would we know? We've been told Darwinian evolution is a fact. That, due to *random* genetic typos, bacteria became worms, spiders or fish. Fish became dinosaurs, dinosaurs became mammals, mammals became apes, and apes became humans—all due to pure chance. No intelligent input at any stage.

'It may seem ridiculous to those who take a little more time to think about these things, but many probably just take the view that, "Who are we to argue with '**THE SCIENCE**' a that tells us that Darwinian evolution is just a "fact," like gravity. The same science that has brought us so many benefits in other, more practical, fields. We are, I think, in a collective trance around it all, where the "high priests" of this **ATHEISTIC SCIENCE**, which has brought us so many practical benefits, insist that the way they see reality, that is as *soulless* and *spiritless,* without meaning or design is the right way.'

'Ironically, Darwin, himself, conceded that,

> "The number of intermediate varieties, which have formerly existed on the earth, **(must) be truly enormous**. Why then is not *every* geological formation and every stratum *full* of such intermediate links? ['*Every?*' '*Full?*' **There are none!**] Geology assuredly does not reveal any such finely graduated organic chain;" [Oh, dear, **it doesn't reveal any such chain *at all!***] "and this, perhaps, is the MOST OBVIOUS AND GRAVEST OBJECTION which can be urged against my theory." 143 [Ah, finally. . .]

'Because, far from providing such a "finely graduated chain" and an "inconceivably great" number of links "between all living and extinct species," the fossils provide no genuine links at all.

'Darwin's technique with objections, was to make excuses as to why—despite the fossils and the data from embryology not agreeing with him—his theory *'ought'* to be right because, he felt, it was such a good one.

'But didn't he say, we need to reject his theory, unless we *truly* think that the **lack** of transitional fossils, the **sudden,** not gradual, species appearances and the **limited** powers of natural selection all don't matter?'

'Yes, but how sincere, really, was he when he said that? Because, it seems that he felt his theory was so good that, surely, it had to be right—didn't it?

'So, rather than testing it continuously against the data, he and his followers have, consistently, tried to make the data fit with his ideas.'

'As we saw with Eldredge and Gould's theory of punctuated equilibrium?'

'Yes. We do not need to disagree that Darwin had an interesting idea. An idea which needed to be tested. Amazing if he was right. How did Life unfold?

'Well, "Once upon a time there was a tiny, non-nucleated, single-celled, bacterium and it varied by chance, and it varied and varied until it became a worm, or spider or snail, or fish, then a dinosaur, bear, and a whale!"

'Ha ha Ollie. But, surely people were right to consider Darwin's theory?'

'Yes, but, as I said, it needed to be tested: Could life arise by pure chance? Could the random mixing of NON-LIVING CHEMICALS *really* create *Cells and Shells, Wings and Feathers!, Livers and Kidneys, Hearts! and Lungs, Muscles and Sinews, Hands and Fingers, Eyes and Brains?*

'Would chemistry, genetics, and, especially, the **fossil evidence**, eventually, confirm his theory? Did Life on Earth unfold as he predicted? In ***Darwin's Enigma***, Luther Sunderland reminded us that the fossils frustrated Darwin, "because they **did not reveal** *any* evidence of a gradual… evolution of life from a common ancestor, proof **which he needed to support his theory**." Darwin just hoped that "future explorations would … ultimately prove his theory correct." Meanwhile, he wrote, by way of 'explanation,'

> "The geological record is *extremely imperfect* and this fact will to a large extent explain why ***we do not find*** intermediate varieties, connecting together all the extinct and existing forms of life by the finest graduated steps ['**<u>finest graduated steps</u>**?' Come on Darwin, we do not find any such fossils at all!]. HE WHO REJECTS THESE VIEWS on the [he said, very poor] nature of the geological record, WILL RIGHTLY REJECT MY WHOLE THEORY." [144] [At last, reality.]

'Now, after decades of "geological exploration… the picture is infinitely more… complete" than in Darwin's day, our museums are now "filled with over 100 million fossils of 250,000 different species." So, Sunderland asks,

> "What is the picture that the fossils have given us? **Do they reveal a continuous progression** connecting all organisms to a **common ancestor?** With every geological formation explored and every fossil classified it has become apparent that these, **the** ONLY

direct **scientific evidences** relating to the history of life, **still do not provide** any of the evidence for which Darwin so fervently longed. ... [this] **can no longer be ignored or rationalized away** [as Darwin urged, and his followers *still* do] with appeals to the imperfection of the fossil record." [145] [Ah, dear Reader, we do so need to let Darwin go!]

'Unfortunately, for many of us, Darwin's theory has become its own kind of pseudo-religion, which cannot be questioned, without accusations of being *unscientific,* or a scriptural literalist, when neither may be true.

'Lyn Margulis, a well known biologist, agreed that Darwin did not successfully explain natural history. She wrote, '**Natural selection eliminates and maybe maintains,** BUT IT DOESN'T CREATE.'[146] She continued,

'. . . the Darwinian claim to explain all of evolution is a popular **half-truth** whose lack of explicative power is compensated for only by the **religious ferocity** of its rhetoric. [Yes.] Although random mutations influenced the course of evolution, their influence was mainly by **loss, alteration,** and **refinement**. One mutation confers resistance to malaria but also makes happy blood cells into the deficient oxygen carriers of **sickle cell anemics**. Another converts a gorgeous newborn into a **cystic fibrosis patient** or a victim of early onset **diabetes**. One mutation causes a flighty red-eyed fruit fly to fail to take wing. Never, however, did that one mutation make a wing, a fruit, a woody stem, or a claw appear. Mutations, in summary, tend to induce **sickness, death or deficiencies**.

'**No evidence** in the vast literature of heredity changes **shows unambiguous evidence** that random mutation itself, [RGMs are genetic typos,] even with geographical isolation of populations, leads to speciation. Then *how* do new species come into being?'[147]

'In an interview she said,

"This is the issue I have with neo-Darwinists: They teach that what is generating novelty is the **accumulation of random [genetic typo] mutations in DNA, in a direction set by natural selection**. If you want bigger eggs, you keep selecting the hens that are laying the biggest eggs, and you get bigger and bigger eggs.

"But you also get hens with defective feathers and wobbly legs. Natural selection eliminates and maybe maintains, but it DOESN'T CREATE ... neo-Darwinists say that new species emerge when mutations [genetic typos] occur and modify an organism. **I was taught over and over again** that the accumulation of random mutations led to evolutionary change – led to new species. I BELIEVED IT UNTIL I LOOKED FOR EVIDENCE." [148]

'Doubts about Darwin are nothing new. **Darwin had his doubts**. He just hoped that, in the end, he'd be proved right. But, his claims for the powers of *natural selection* to create and to evolve all Life, were simply assertions as to what he thought *might* be true, *not* the evidence he needed to show that they *were* true. Hence, he *bluffed and bluffed,* always promising that he would prove his claims in a later work, which, it's no surprise, he never published. A famous modern follower of Darwin, Ernst Mayer, wrote,

> 'It must be admitted, however, that it is **a considerable strain on one's credulity** to assume that such finely balanced systems such as certain sense organs (the eye of vertebrates, or the bird's feather [or lungs, or livers, or our senses of smell and hearing, or blood, or hearts or brains]) could be improved by random mutations.'[149]

'That sounds like a massive understatement to me, Ollie!'

'I agree. Malcolm Muggeridge, a well known 20th century intellectual, observed that Darwin's "theory of evolution," which we have been taught, by our scientific "high priests," is *scientific gospel,* and not to be questioned, "will be one of the greatest jokes in the history books of the future. Posterity," he continued, "will marvel that so very flimsy and dubious an hypothesis could be accepted with the incredible credulity it has."[150]

'Darwin, and others of his era, wanted something more scientific than: "Read the Bible, Creation is described in there..." To this extent they were right. The scriptures are not *'science.'* They are just *poetic* symbolizations of the classical inference that *Life and Mind, our Love of Goodness, Beauty and*

Truth, our Intelligence, our Curiosity, and our Creativity's ultimate sources are **meaningful** and **intelligent,** not MINDLESS, not RANDOM—that's all.

'Darwin, and some of his contemporaries, went to the extreme of trying to find totally *non-spiritual,* **non-intelligent** explanations for *Life,* and *Mind.*'

'Which, *if* it worked, would be fine... because it would be the truth.'

'Yes, if it worked. Unfortunately, Darwin, and so many since he, seem to conflate the validity, or otherwise, of traditional religious views with the more basic question as to whether the *super* natural exists at all. Whereas, I think the *super* natural is real, demonstrably so, but that the religions, while they have the basic fact of the existence of the *super* natural right, tend to get many of the details wrong, bizarre ideas around heaven and hell, for example, which is why many, today, will have nothing to do with them.

'Too often they have portrayed *god or spirit* as this grim, punitive force. Is it any wonder then that people are no longer drawn to their teachings? They want us to believe in their dogmas, but take no notice of the modern *psychic* and *spiritualist* research which challenges some of their dogmas.

'This, though, should not stop *us* from knowing that there is far more to Life on our planet than thinkers like Darwin are willing to allow. There is, clearly, a *vast* intelligence in the *moment to moment,* arising of living things —each of us breathing, our blood circulating, trillions of biochemical events taking place in our bodies, *right now!, this second!, every second!*

'We need to *follow* the data, not to keep fighting it, as Darwin's followers urge us to do. Like the well known biologist, Richard Lewontin, who, so strangely, and so sadly, in my view, writes,

> "Our willingness to accept scientific claims that are **against common sense**[!] is the key to an understanding of the real struggle between science and the **supernatural** *[intelligent design, spirit].* We take the side of science in spite of the **patent absurdity** of some of its constructs..." ['Life is just a chemical fluke'] "in spite of the tolerance of the scientific community for *unsubstantiated just-so stories* [like Darwin's scheme], because we have a ***prior commitment,*** a commitment to MATERIALISM" [151] [to a spiritless, dead, universe, to MERE CHANCE as a supreme creative principle—at least he's honest] ...

"Moreover, that materialism is absolute, for ***we cannot allow*** a Divine Foot [intelligent causation] in the door. ...To appeal to an omnipotent deity is to allow that at any moment the regularities [laws] of nature may be ruptured, that Miracles may happen."

'Yet, science, Ollie, can only be in a struggle with the supernatural, as he puts it, if we know the latter to be unreal—which we don't.

'Surely by denying all evidence for ***intelligence*** in nature, including, to be consistent, in we, who are a part of nature, Lewontin is asking us to put science into a struggle, not merely with the *spiritual,* but with ***reality itself,*** with how things *really are.* I don't understand how our ancient *knowing* that all living things arise from subtle *spiritual* sources became so taboo?'

'Well, prior to Darwin, it was accepted that many things, especially *Living things,* had facets beyond their purely PHYSICAL components—like the *Life Forces,* and the *Soul.* It was what most of the world's great thinkers had always held. It was only during the 19th century that the *dismal spirit of* ***materialism***—what the great Austrian philosopher, Rudolf Steiner, referred to as *Ahriman*—gained such a grim sway over the western mind.'

'I'm curious, though,' said Alisha, 'that it doesn't occur to Lewontin that the *laws of nature,* which he so values, can stem from a mysterious cosmic intelligence *which 'just Is,'* no less than the *mindless* matter, and its laws, which he believes in, *'just Is.'*

'Nor does he allow that, far from being violated by this *subtle, cosmic, intelligence,* which ***just Is***, the amazingly elegant astonishingly precise laws of nature are *generated by* this *same intelligence,* the logical and necessary existence of which he so dismisses, such is his *faith* in the powers of blind chance, which *'just Is,'* to create and to evolve all.

'Also, while commonsense is not always a good guide, to rule it out of our scientific thinking altogether would be odd, wouldn't it?

'The classical, commonsense, hypothesis—which Lewontin rejects—is that neither Life nor Mind could *possibly* arise by chance, they are far too complex for that.'

It was time a for a rest. We sat in the shade, by the sparkling river.

14 Science *vs* Materialism

'One question, Ollie. Why *is* science, today, so opposed to all ideas of *non-physical* causation? Why the relentless denial of *Soul and Spirit?*'

'Because Life's *deeper* causes cannot be seen? Because we can only know them *indirectly,* or *infer* them, using logic. Most academics today seem to fall into the materialist camp, so even if Darwin is wrong, it will make little difference to their overall view. It would, therefore, need a big shift for them to reopen their minds—*and hearts*—once again, to any ideas of *spiritual* causality. A shift which, many think, would be damaging to science, as we know it.'

'Why?'

'Because, currently, to allow any ideas of *super* natural causality into our understandings of nature conflicts with the science method conventions known as method.o.logical materialism or naturalism (MM/MN)—two ways to say the same thing. According to MM/MN scientists are to assume—*even if* the *spiritual* is real—that there is no intelligence or design at any stage, that MINDLESS CHEMICAL CHANCE can create all—genes, DNA, *Cells, Blood, Hearts, Lungs, Brains,* everything—for no reason at all.'

'That's crazy, Ollie. We know perfectly well that mere chance cannot create anything remotely clever! Surely this approach makes it impossible to *fully* explain so many things in Existence: *Living Beings, Mind, Love, Will, Curiosity, Creativity,* let alone *C O N S C I O U S N E S S itself...*'

'I agree, but the original idea of disregarding two of Aristotle's four causes, the *shaping ideas* and the *purposes* of things, and of assuming that only MATERIAL and MECHANICAL causes are operative in nature, did have *some* benefits. It did help to steer scientists away from the medieval tendency to turn to *super* naturalist explanations whenever it seemed difficult to explain something in purely physical terms.

'This approach, adopted at the dawn of the modern age by thinkers like Sir Francis Bacon, would, it was thought, reduce the temptation to indulge in what some now refer to as '**god-of-the-gaps**' thinking, where, if scientists don't understand something, they may try to fill a knowledge gap by saying, "this may be *supernatural* in origin, so we won't be able to make *scientific* sense of it. We must just accept this and move on."

'Hold on, Ollie. Just because something has an *intelligent* cause, it doesn't mean we'll never be able to understand it, do good science on it, or harness it for practical purposes. After all, can't MM/MN lead, just as easily, to lucky '**chemical-flukes-of-the-gaps**' thinking? Where, every time scientists don't know how something arises, like a DNA CODE or Cell—perhaps by extremely clever processes—they try to fill their knowledge gaps by saying,

> "This complex mechanism **must have** arisen by chance. It may *look 'designed'* but it's an illusion. We can't PHYSICALLY see any intelligence(s) behind it, so it *must* be ra*N*dom in origin."

'Well, it is true that whether we posit *intelligent* causes, or MINDLESS, we are trying to fill a knowledge G A P. We just have to decide which type of explanation better fills the gap. A landscape evolves randomly, by unintelligent design or **UD**. But Cars, PCs and TVs only make sense in terms of intelligent design or **ID**. Logically, the same reasoning applies to Living things—all far more complex than anything we make.

'I agree, Ollie. So what is the benefit of MM/MN, especially when it comes to our understanding of Living things?'

'On the whole I agree. But MM/MN does have *some* valid points.'

'Like?'

'Well, take thunder and lightning. For a long time they were thought of as wholly *super* natural phenomena. But, eventually, scientists found they could be described quite well in physical science terms.

'Similarly for magnetism and electricity. If scientists had given up and said, "These are supernatural things, we'll never understand them," we would not now have the many practical benefits our modern understanding of them has made possible, TVs, Cars, and PCs for example.'

'Hold on, Ollie. Just because we understand electricity or magnetism better today, it doesn't prove that their origins lie in the mindless, does it?'

'I agree. In any case, the distinctions we draw between the *'everyday'* natural and the *'super'* natural are of our own making. They are simply our labels for what we think **Exists** or **Is**. *What Is* is, ultimately, a ***whole***, be it solely physical or *spiritual* as well, and what is more *'natural'* than *'that?'*

'Whether we label some of *'**What Is**'* as *'ordinary'* natural, and some of it as *'super'* natural, we can attempt to investigate *all* of it—because we want to understand the ***totality*** *of Nature, of **what Is**,* physical and *non-physical*.

'This is the debate: what is the true nature of *Nature, of what Exists?* Are the materialists correct, or the *spiritualists?* What is the true nature of *'What Is,' of All that Exists?* Is only the physical real? Or is there more to reality than that? Is reality MONO-DIMENSIONAL or *multi-dimensional?*

'Unlike Darwin, we now *know* that Life could ***never*** arise by chance. The only reason that some of those, like the distinguished chemist, George Wald, who, having admitted this same point, then reject their own logical conclusions, is that it falsifies their *philosophy*. Wald wrote:

> 'When it comes to the origin of life there are ONLY TWO possibilities: CREATION [intelligent causation] *or* spontaneous generation. There is NO THIRD WAY. Spontaneous generation was disproved one hundred years ago, but that leads us to only one other conclusion, that of *supernatural* creation. We cannot accept that on **philosophical** grounds; therefore, we choose to believe the I M P O S S I B L E : That life arose spontaneously by chance!'[152]

'This is not, as Wald acknowledges, a strictly scientific attitude, but a philosophical one. The trouble is, materialism forces us to look for NON-SPIRITUAL explanations for everything—even when it doesn't work—even when trying to explain *Life, Mind, Intelligence, Feelings, Desires, Love.*

'Just because we cannot yet work out exactly how *(or why)* an *underlying* intelligence(s) or *Higher Power(s)* give(s) rise to *Living, feeling, creating, Human Beings* like us, does not mean that there is, or there are, no such *force(s),* nor that we cannot work out anything, at all, about *It,* or them.

'None of us **know** that an *inherent, Subtle, All-Pervading, Creative, Cosmic Intelligence* does not exist. Our materialist friends *think* it does not, because they cannot PHYSICALLY see it, and because they are unwilling to *infer* it. What they neglect, though, is that **'intelligence,'** as a `quality,` is not something that can *ever* be seen—at least not physically.'

'Go on. . .'

'Because it is a *spiritual* quality. We can only see its *effects*. If they were alive today, we would not be able to see Plato, or Galileo, or Newton's *intelligences,* only their writings and their other creations.

'Even if the *intelligent, spiritual Source or sources* of living things were physical beings, like us, we would not be able to see their *intelligences,* only their creations.

'We cannot *physically see* the incredible intelligence that goes into the *24/7* functioning of cells, vastly cleverer than anything we make—only its *effects*. We cannot *physically see* the *source* of the intelligence of the design of the eagle's wing, the peacock's tail, or the bat's sonar—only its *effects*.

'We cannot even *physically* see the *intelligences* of the clever people who make our Cars, TVs or PCs, we can only see the *results* of their intelligences.'

'But we can see their BODIES, their hands, their faces, can't we . . ?'

'Yes, but we cannot examine someone's BRAIN CHEMICALS and find their *intelligence*. The same with *spiritual* intelligence. Physically, we can see only its *effects*—all around us, all the time, as it happens.'

'In any case, Ollie, to say that nature is *not* intelligent, or *not* purposeful, including we, who are part of nature, seems so strange.' Mused Alisha. 'Yet, once we admit to intelligence *in* nature, we have to admit to reason and purpose *behind* nature, because intelligence always has a ***source***, with reasons and *purposes,* which are always motivated by intentions and desires —by what is *loved*. But only *subjective Being or Beings* can love. Only *It, or they,* can have intentions and desires—this points *to Spirit or Spirits.*'

'I agree. But regarding the science methods which encourage this way of thinking, the issue is that until not so long ago, if scientists didn't understand something, like electricity or magnetism, they saw it as magical and mysterious. But, once they had understood it, well enough to use it, instead of continuing to view it as yet another *miracle* of a basic reality which is, in its very nature, miraculous, it was demoted from the *'super'* to the, 'it's-nothing-so-special' natural, forgetting that the fact that *any-thing-at-all exists,* vs. nothing-at-all, is the most *extra-ordinary,* the most *'super,'* or the most *astonishing* thing of all—because the impossible has happened . . .'

'You said that before. . .'

'Yes, that there is *any.thing* at all. How is it possible? Shouldn't it be *im,possible?* How is it possible that there is *any.thing* rather than *no.thing?* Be it *Spirit-Being? Or Mind? Or natural laws? Or Big Bang? Or atoms? Or stars? Or cells? Or blood, or rivers, or trees, or people,* vs nothing at all? *Where* do all these astonishing things *come from?*

'Isn't it easier, Ali, to imagine there being *no-thing* at all, rather than some-thing at all? How can there be anything? *Where* can it come from?'

'Well, everything comes from the *Big Bang,* doesn't it?'

'But what caused the Big Bang? Something must have… something that *'just Was'* and, presumably, *still Is.* The Big Bang doesn't explain its own cause. It doesn't explain the fact of there being a Cause of anything-at-all, rather than nothing-at-all, does it? How come this (Uncaused?) Cause exists? **How** can it exist? **Where** would it come from? No one can say.

'But, what we **can** ask is this: what is the *True Nature* of this: *"It may seem impossible that it just Exists, yet **It** just does, Source-and-Cause-of All?"* Is it mindless or *intelligent?* If the Source of Existence is MINDLESS, where do our *minds* come from? Which of the two *'impossibles'* makes the best sense, overall? MINDLESS ultimate source of all *or* Intelligent?

'Culturally, because we rejected the too literalistic beliefs of some religious people, we opted for a view of reality, the MATERIALIST view, which, although it is not totally illogical, leaves so much more that it cannot explain. For example, is it genuine *science* or is it ideology to dismiss every single psychic, mystic, shaman or sage who has said that *Being* consists not just in MINDLESS MATTER *but also in* **Life-Mind-Sentiency**, in *Mind and Soul* realms, just because materialism cannot accommodate their insights?'

'The spiritualists and the shamans, the sufis and the yogis have always said that reality is *multi-dimensional,* that it encompasses *Subjective,*

Spiritual elements, which can be encountered by *inner* means, by shifts in our *own consciousnesses* as we travel on the various spiritual paths.

'Currently, though, our sciences don't see it this way. As each natural miracle, like magnetism, electricity or genetics, has been explored, and put to good *(or bad)* use, it has gone from being thought of as *super* natural to, merely, *'ordinary'* natural. A process which is called *naturalization*.

'Eventually, scientists decided that this same process might happen in relation to everything they set out to investigate, *Life, Mind, psychic* phenomena of all kinds. All things would be *'naturalized,'* that is, explained in purely physical terms—and no cause *beyond* physical nature would ever be needed to explain anything. A flawed way of thinking which leads to < < < <   THE GREAT ERROR   > > > > Because, to use the skills of modern, physical science to clarify Aristotle's first two causes, what things are '**made of**' and '**how they work**,' and put them to use, clever as it is does not tell us *how, or why,* they arose in the first place.'

'You mean did they have random causes, or meaningful?'

'Yes. Physical science, by itself, can tell us nothing of Aristotle's *formal* and *final* causes. Cleverly investigating made of *and* how it works, the material and mechanical sides of nature's holistic 'coin,' does not tell us of their *deeper, meta-physical* causes, nor anything about the *inner,* `spiritual` dimensions of reality. Some say this doesn't matter: "Who cares whether there is anything meaningful *beyond* obvious, OUTER nature, or any continuity of *Mind,* or "`I-Am-ness,`" or links with our loved ones after death?"

'But those are all interesting things to explore, aren't they? They concern some of the deepest things we can think about.' Sighed Alisha.

'Yes. Science is about finding out what is *true*. The *latin* word, *'scire,'* from which our word, *science,* comes, doesn't mean materialism. It means **to know**. So, is it *true* that all things in nature are just coincidences?

'Using a purely materialistic methodology in science started out as a (sometimes) useful idea. But, in time, it morphed into the dogma that there are **no** phenomena which do not have purely materialistic explanations, and the claim that there are, probably, no *spiritual* components to reality at all.

'This is how the views of materialists like Darwin, Marx, and Lenin, became more fashionable than those of *spiritualists* like Plato (theory of Ideas, Platonic Forms), Galileo (heliocentricity), Newton (laws of motion, gravity, *Principia Mathematica*), Faraday (electromagnetism, electro-chemistry), Maxwell (mathematical-physics, electromagnetic-radiation), Planck (quantum physics) and Lodge (radio, spark plugs).

'This is why it became unfashionable for scientists to investigate phenomena like *telepathy* and *mind-beyond-the-body,* and why more open minded thinkers like Behe, Meyer, and Dembski, who put forward powerful arguments for **intelligent** design, have been dismissed or ignored.

'Unfortunately, under the materialist paradigm of **unintelligent** design, any scientists who wonder whether something may *not* be susceptible to a purely physical explanation are *treated like heretics,* and urged to confess, quite unfairly, that they would not be clever enough to detect any evidences for intelligence or design in nature—even if they are staring us in the face.

[I hope, dear reader, that one day it will seem utterly ridiculous, that authors, like myself, have had to devote so much time to pointing out the utterly obvious, while our materialist and atheistic friends, like Dawkins, so blithely, yet so unreasonably, claim the intellectual high ground—insisting that 'science' is on the side of materialism and Darwin's theory of 'unintelligent design!' Ah me!]

'Take *psychic* abilities and *super* natural experiences of various kinds. Conventional scientists have never been able to find purely physical explanations for such things. So, the tendency is to ignore or to deny them[153] or to accuse colleagues, like Rupert Sheldrake, curious enough to research them, of being deluded, susceptible to hoaxes, or fraudulent.[154]'

'Yet if something,' started Alisha, 'like a DNA Code, a Cell, an Eye, or the Human Mind, *does have subtle, non-physical* causes, is it right to argue that we can explain it *as if* it doesn't? A people confronted with a never-before-seen mechanism, like a Car, TV, or PC, not seeing the clever intelligences who made the Car, TV or PC, could hypothesize that the unknown mechanism may have arisen due to purely natural processes…'

'Yes, they could take it apart, randomly, tumble its parts around, and see if it would *'naturally'* re-assemble into its original, purposeful sequences of parts, due to the natural laws of physics, in combination with the natural laws of random tumbling. ***Would this work?!*** Such experiments have been tried with *living* things, like bacteria, random swirling of the parts carried out, electrical sparks fired, and no new life forms have *'naturally'* emerged.

'However, if our 'philosophy' does not allow us to see all the utterly ***obvious*** evidences for intelligence and purposeful functioning in nature—just looking in the mirror *ought* to be enough—this will make no difference.

'Like modern, versions of poor Sisyphus, we will futilely continue to try to explain *all* the data of reality, even our own, *staggeringly* complex bodies, even our own *bright* minds, in terms of entirely mindless forces.

'Some of us may even claim to find this sterile, *inquiry-limiting* way of thinking 'intellectually fulfilling!' It becomes our grim *task,* and our dreary *faith,* which we have persuaded so many others to believe in, just because we have convinced ourselves that only the obvious, SOLID, MATERIAL UNIVERSE can be real. No, `subtle, spiritual, causes,` are allowed in this dreary ontology of mindless, random, pointlessness . . .

'Talking of *faith,* Ollie, what about the religious? Where do they stand on all this? Plenty of religious people also believe in Darwin, don't they?'

'Well, some try to follow Darwin, minus his MATERIALISM, some not. But, for the religious, it can be complicated. On the one hand, they believe in the *super* natural. On the other hand, many religious people believe that science cannot tell us anything about the non-physical aspects of reality.

'Why not?'

"Because science," they say, "can only concern itself with the PHYSICAL." As to *super* natural or *spiritual* matters, "Only religion," they feel, "can give us any answers." Answers which we must either accept on *faith,* or reject. Not all spiritual people see it this way. Modern theorists of intelligent design, like Behe, Meyer, Dembski, do not claim that science can confirm their own or anyone else's private, spiritual convictions. But they do assert that today's scientists are **perfectly capable** of working out whether, or not, there is more to Life than just the physical, and whether, or not, certain features of nature are more likely to have *intelligent* causes than random.'

'So what's your conclusion, Ollie?' Mused Alisha.

'That science does ***not*** *'have to'* confine itself to the material alone. Sure, some things in nature, in the totality of ***what Is,*** are easier to investigate than others. But we do not need to restrict ourselves to investigating the **PHYSICAL** alone, either (a) because **some religions** want to *frighten us away* from knowing anything more about the *non-physical* than *they* believe, or would like *us* to know, or (b) because **materialists** refuse to accept that there can be anything *beyond* physical nature to inquire into.

'My view, Ali, is that although *many* things can be explained quite well in purely physical terms, not *all* can be. Trying to reduce *clever mind* to **MATTER** alone being one. Insisting that Life's **CODES**, *Cells!,* amazing organs like *Blood! and Hearts!, Kidneys! and Ears, Eyes and Brains!, Hands and Feet,* could form by blind chemical *'luck!,'* being very obvious others.

'I just wish more of us would reopen to the classical idea that, once we have exhausted all reasonable possibilities for a purely physical or MM/MN explanation for something, like a DNA Code, a Cell, a Butterfly, a lovely Dolphin, or a gifted Human Being, why not give space to the classical idea that **Spiritual Intelligence(s)** may be involved?

'This doesn't mean that we have to become ultra religious or *'spiritual'* all of a sudden. Let's just stay open, as a scientific culture, to the thought that there may be ***more*** to our existences than the purely physical. That, as the *spiritualists,* prophets, psychics, mystics, yogis and shamans, have always said, reality is *multi-dimensional,* not **MONO**. A dogmatic materialism, like Lewontin's, unduly constrains science. It says, "What's the point of trying to investigate the *non-physical?* We assume, as scientists, that it does not exist."

'That sounds incurious, and lazy, to me, Ollie. What if there *are* non-physical components to reality which *can* be investigated, at least a little?'

'I agree. Darwin's famous colleague, Alfred Russell Wallace, did precisely this, and *he* concluded that, "the phenomena of *Spiritualism* ... are proved quite as well as facts are proved in other sciences." [155]

'What if reports of the *multi-dimensional* nature of reality, like Wallace's are *true?* ***Are we to reject*** all such evidence, and to dismiss every single *spiritual, religious, or mystical* experience anyone has ever had, just to try to make materialism's gloomy, reality-shrinking, claims seem true?

'Must we ignore every interesting coincidence, NDE, OBE or other *super* natural experience anyone has ever had, just because materialism cannot accommodate this kind of data? Materialism is a set of beliefs about the *basic nature* of reality which has huge spiritual implications for us.

'It tells us, "Sorry, there is nothing deeper to reality than blind chance." 'It tells us, "Don't be fooled, there is no deeper wisdom you can tune in to, if you are prepared to self-quieten, and to listen." No, "We are all just pointless accumulations of CHEMICALS, nothing more, without *purpose* or *Soul."*

'It insists that *this* world is all there is, and that, when the 'accidental chemicals' stop, we drop, and that's it. With such grim implications as these, 'materialism' needs to be tested, and tested hard. Then, if it is found wanting, it may be that there are other, better ways of making sense of existence. A key test, by the way, is that 'materialism,' as a philosophy, is self-refuting.'

'Go on...'

'Because, Ali, it is a *belief* system. It consists in A B S T R A C T I D E A S! But I D E A S are not physical things, are they? They only exist in the *non-physical realm of the Mind. I D E A S* are not made of PHYSICAL stuff. They can be *symbolized by* **matter**, using **ink**, for example, just as the ideas expressed in Life's CODES can be symbolized by amino acids, but, of themselves, they are immaterial, *abstract, informational.*

'Even more ironic, in deep reality, matter, as such, does not ***truly exist...*** as Max Planck, brilliant discoverer of quantum physics explained, "THERE IS NO MATTER AS SUCH. All matter originates and exists only by virtue of *a force. . .*mind is the matrix of all matter."[156] He discovered that our seemingly so SOLID physical world is not ultimately 'S O L I D' at all. Actually, it is 'made' of S . . . P. . . A. . . C. . . E . . . 9 9. 9. . .9. . .9 . . .9 . . .9 . . .9 . . .9 . . . 9. . . 9. . . 9. . . 9. . . 9. . .% of it.

'So that makes 'materialism,' as an ideology, self-contradictory?'

'Yes, deeply. Another interesting thing about us humans, Ali, versus, say, worms, or dogs, is that our human needs go far beyond those needed just to *live*. Yet, the ***cells in a worm's body*** *are **just as complex*** as those in ours. Their cellular needs are no less than the needs of our cells. Isn't that curious? Yet, surely, it is safe for us to say that a worm's consciousness, and its requirements for satisfaction, are vastly more limited than ours. Yet why, on a biological level, would the cells in a worm's body need to 'evolve' beyond their worm-based stage? Surely the cells in a worms body are just as contented as the cells in our bodies. Yet, *something* causes the different Life Forms to exist—with very different levels of C O N S C I O U S N E S S.

'At its own level, be it part of a worm, fish, dog, bird, or human, *the living Cell is the **most complex mechanism*** known to us. Yet, paradoxically, it is the ***most basic unit*** of all living things. So evolution, in so far as it is a facet of reality, may be less about different levels of bio-chemical organization, and more about different levels of C O N S C I O U S N E S S.

'For example, while dogs require much more than worms, and share many intriguing and fun qualities with us, once their fairly simple needs are met—food, shelter, walks, friendly contact and play—they are happy.

'Yet, our human desires and our human capabilities take us *far beyond* any logical biological needs—like being well fed, sheltered, and safe.

'Are you saying, then, Ollie, that if a totally *mindless* form of evolution, like Darwin's, were true, why would life ever 'purposelessly' and 'soullessly' need to evolve beyond some planet covering algal or bacterial stage? Why would it need to? It would be the most successful Life Form.'

'Yes, exactly. Dawkins has argued, in his famous ***'selfish-gene'*** theory, that ***genes*** are the true beneficiaries of natural selection—that evolution is nothing more than a vehicle for genes to perpetuate their own existences.

'But if the only *'purpose'* of evolution is to perpetuate genes, there is no reason for such a ***'selfish'*** kind of Life to ***'evolve'*** beyond some planet covering algal or bacterial stage. In fact, ***bacteria*** may be the most successful life forms. So why would they ever need to *"selfish-gene"*[157] evolve into anything else—quite apart from the fact that there is ***no evidence*** that they ever did so.[158]

'Not only is there ***no extraordinary evidence*** for the quite extraordinary idea that ***random, genetic copying error, mutations*** could create, then convert, a basic life form, like algae or bacteria into worms, or snails, or spiders, or fish, or birds, or dogs, let alone, humans, it **doesn't make sense** . . .'

'Yet science, surely, should always make sense, Ollie, shouldn't it?'

'Of course it should. Which, is why it needs to be rescued from materialism! For example, unlike worms or dogs, we, humans, crave meaning and truth. ***Why?*** *What* is it in us which can *look* and *listen* and *care?*

'**What** is it in us which *cares* about the very *nature* and the very *truth* of things? *Why* are thinkers like Darwin and Dawkins so passionate about such immaterial, spiritual, qualities as **truth?**

'*What* is it in us which can *remember* experience? And *reflect* on it?

'*What* is it in us which can *imagine* and solve abstract problems?

'*What* is in it us which *cares and loves* and grieves and rejoices?

'*What* is it in us which can *imagine* Cars, PCs or TVs into existence?

'Is all this *really* just due to MINDLESS ATOMS OF CARBON, HYDROGEN and OXYGEN, having rAn*d*0*m*ly bumped and clumped together, for no reason at all, and, eventually, become us? Is this *honestly* intellectually satisfactory?

'Or is it the *Soul* and the *Spirit* in us, as classically understood? We, as humans, are almost *god-like*. If all we were made of was 'DUMB' MATTER + DUMB LUCK, how could we do all these amazing things, let alone care?

'Yet, despite all the pseudo-scientific *puff* pushed at us, by Darwin's fans and other materialists in the origins debate, some deep impulse in us does want to explore ever deeper into the *mysteries of existence*. Firstly, the outer worlds of PHYSICAL FORM.

'How is a cell or worm or dog put together? What is it **made** of? What is an atom, planet or star made of? How does it **work?**'

'But cleverly working out Aristotle's first two causes, *what* things are **made of** and how they **work,** does not tell us *how, or why,* they arose in the first place, does it, Ollie? Nor, why they arise right now, *this* amazing second, and every amazing second, 24/7. What are their *deeper* causes?'

'While most of us cannot see the *inner plane* dimensions, the *spiritual realms,* the world's shamans and *mystics, its yogis* and its psychics tell us that not only can we infer *Life's deeper causes,* we can also do this from *within.*

'They say that, by **going inwards** in meditation, visualization, and prayer, we can, not only ***deepen our contact*** with our own *Souls*, we can, potentially, achieve ***some measure of direct contact*** with the deeper sources of things, which ***our exclusively*** OUTER directed inquiries will never show us.'

We sat quietly for a while, watching the stream and eating our snacks.

15 Icons of Evolution or Fakes of Evolution?

'We have been told, repeatedly, that Darwinian evolution is a fact. However, in *Icons of Evolution: Science or Myth?*, Jonathan Wells examines the surprisingly few items Darwinians have had to try to prop up their shaky theory, and he shows how they are all, either MISLEADING, or false or fake.

'Go on...'

'We have already met some of the dodgy icons. Miller and Urey's famous 1950s origins of life experiments, which many were led to believe were proof that Life might arise by chance, when all the experiments produced were a few amino acids. Well, it's true that amino acids are among the basic building blocks of life, but building blocks, unguided by higher level forces, will never build anything functional.

'In *Evolution's Final Days*, John Morrison asks, "If we threw a pile of blocks on the ground would they land in an ordered pile? Of course not. We could throw the same blocks for ever. We would never get an ordered pile."

'It's obvious . . . but spell it out, Ollie . . . why not?'

'The laws of probability. Morrison asks: "Why are the laws of probability taken into account in all other areas of science, but *not* when it comes to evolution? The **theory of evolution states that random amino acids somehow came together**, thus violating the 2nd law of thermodynamics [entropy] and became a [very complex *living*] cell. Then, this single celled, asexual organism randomly mutated and... has since **become [ever] more ordered** with every mutation."

'Darwinians appeal to vast time to justify this idea, yet as Michael Denton says, "**time in itself tells us nothing of the probability of achieving any sort of goal** unless the complexity of the search can be *quantified.*" [159] Darwin did not even try. How come? Why did he not try?'

'The next icon Wells considers is Darwin's famous tree of evolution. If biology textbooks were to be more candid, and more accurate, they would explain that Darwin's notion of microbe, to vast, spreading, tree of Life remains unproven. In fact, some Darwinians are starting to say that the evolution they now believe in is not tree-like at all, but network-like.

'But that's nothing like Darwin's concept of evolution, is it? Because, in a network or web-of-life, there is no discernible start or progression...'

'Correct. Another famous Darwinian icon relates to homology.'

'Homology?'

'Homology refers to the interesting similarities in skeletal structures shared by many vertebrates. For example, the similar forelimbs of bats, porpoises, horses and people, including the common five finger pattern. Prior to Darwin, these intriguing similarities were seen as design archetypes of a *spiritual* intelligence which *just Is*.

'Darwin, argued for a different idea which, he hoped, the data from the fossils and embryology would, in time, confirm. What if, he asked, *Living* things were *not* generated by *spiritual* intelligence? Why not *imagine* microbial life arising, by chance, and everything else coming from that. But: (i) This did not *actually* explain how *Life* started. (ii) **Not even a few** (let alone *huge* numbers of) linking fossils are ever to be found.[160] (iii) **There is no,** non-intelligent mechanism which could drive the process Darwin had in mind.

'Yet, Wells says, most text books do not point these things out. Neither do they explain that *both* of the mechanisms proposed to account for these homologies—similar genes and similar developmental pathways following conception—*fail*. They do not point towards common descent but to **separate** lines of descent, which points, more, to intelligent causation, not to rANDoM.

'Darwin did not know of genes, but his modern supporters, who do, claim that these similarities arise, either, (a) due to **similar genes** inherited from a common ancestor, or, (b) due to **similar growth pathways** following conception. *If* it can be *shown,* then, that similar structures in different species **are produced by** similar genes, this would be supportive of Darwin's key idea of a branching, tree-like descent from a universally shared ancestor.

'"But," Wells writes, "**this is not the case** and biologists have known it for decades. In fact, all the way back in 1971 Gavin de Beer wrote:

"Because homology [similarity of form] **implies community of descent** from . . . a common ancestor it might be thought that genetics would provide the key to the problem of homology [skeletal similarities]. This is where the **worst shock of all** is encountered . . . [because] characters controlled by identical genes are not necessarily homologous [similar] ...homologous structures need not be controlled by identical genes."[161]

'Equally striking is the fact that not only do *different* genes give rise to **similar** structures but *similar* genes can give rise to **different** structures.

> 'Geneticists have found that many of the genes required for proper development in fruit flies are similar to genes in mice, sea urchins and even worms. In fact, gene transplant experiments have shown that developmental genes from mice (and humans) can functionally replace their counterparts in flies.' [162]

'Darwin's supporters argue that similar biological structures arise due to a **shared** evolutionary history. If true, this ought to be reflected in similar growth pathways following conception. However, their proposals do not match the data.

'Yet, "It is a familiar fact," wrote embryologist Edmund Wilson, in 1894, "that parts which closely agree in the adult, and are undoubtedly homologous [similar], often DIFFER WIDELY IN LARVAL OR EMBRYONIC ORIGIN either in **mode** of formation or **in position** or **in both**."

'Wells adds, "Gavin de Beer agreed: "The fact is that correspondence between homologous [similar] structures [in the adult forms] **cannot be pressed back to** similarity of position of the cells in the embryo, or of the parts of the egg out of which the structures are ultimately composed, or of developmental mechanisms by which they are formed."[163]

'Yet, Wells notes, most biology textbooks *fail to point out* that the two mechanisms proposed to account for the similarities in vertebrate limbs *do not* support Darwin. So, he asks, if the continuity of information required to build up vertebrate limbs comes *neither* from genes, *nor* developmental pathways following conception, *how* can it come from common descent?

'Moreover, we *haven't a clue* how this complex genetic information arises. I'm ok with that, Ali, because, all things considered, it is quite clear by now that there is more to *Life* than physics and chemistry alone.

'So, although the data points to *separate* lines of descent, this is resisted, because it conflicts with Darwin's ideas of how things ought to be . . ?'

'Unfortunately, yes. But, have you noticed a pattern here?'

'Let me guess… it is always *'the evidence that's wrong,'* never Darwin?'

'Correct! Another dodgy Darwinian 'icon' is Ernst Haeckel's idea that *'ontogeny* recapitulates *phylogeny.'*

'That sounds mysterious,' smiled Alisha. 'What does it mean?'

'It means that the movement of water breathing creatures, out of the sea, onto the land, via an *amphibious,* water-breathing to air-breathing stage, may, even now, be seen recapitulated in the early stages of embryos, including what, at first glance, can appear to look like vestigial gill slits.

'In other words demonstrating a fish ancestry?'

'Yes. To back up this idea, Haeckel produced a set of drawings of different embryos, including fish, salamander, tortoise, chick, rabbit and human, seeking to show how similar they all are—especially early on.

'Unfortunately, to his discredit as a scientist, Haeckel doctored his drawings to make the embryos look more alike than they really are. He did this to bolster Darwin's idea of shared descent. Another (well meant) *bluff.*

'Ironically, what Darwin believed to be the supporting evidence from embryology, as misrepresented to him by Haeckel, gave him more confidence in his theory. He didn't know that Haeckel doctored his drawings, and Darwin said he found them very persuasive. He wrote,

> 'it is probable, from what we know of the embryos of mammals, birds, fishes and reptiles, that these animals are the modified descendants of some ancient progenitor they had in common.'

'Wells explains that since Haeckel's embryo drawings *seem* to "provide such powerful evidence for Darwin's theory . . . some version of them can be found in almost every modern textbook dealing with evolution."

'Yet biologists have known for a long time now that Haeckel *faked* his drawings. Embryos *never* look as similar as he claimed.

> 'Although you might never know it Darwin's "strongest single class of facts" is a classic example of how evidence can be twisted to fit a theory.' [164]

'Wells comments that if one started with the evidence, then examined Darwin's ideas, one would likely conclude that the various classes of vertebrates, rather than being descended from a common ancestor, probably **had *separate* origins**. He notes that Von Baer, a famous 19th century embryologist, couldn't support Darwin precisely *because* of the conflict of Darwin ideas with the real-world data from embryology. Wells continues,

> 'Many modern Darwinists haven't changed. **It doesn't matter how much the embryological evidence conflicts** with [their] evolutionary theory – the theory, it seems, must not be questioned. This is why, **despite repeated disconfirmation**, Haeckel's biogenetic law and [his] faked drawings haven't gone away.'[165]

'S. J. Gould called Haeckel's faked embryo drawings "the academic equivalent of **murder**," a serious charge. Yet, they are still around.'

'So, many people, including other scientists, continue to be misled?'

'Yes, sadly. Wells adds that more accurate textbooks would *not* recycle misleading drawings of embryos, would *not* call pharyngeal pouches 'gill slits,' as though these pouches are evidence of fish ancestry (they have nothing to do with fish gills!), and would explain to students that embryology tends to *refute,* not to confirm, Darwin's ideas.

'Yet, Wells says, many textbooks *still* fail to explain that earlier embryonic stages are **dissimilar**, and they perpetuate the misleading claim that embryology provides evidence for common ancestry.'

'The next dodgy icon Wells discusses is *Archaeopteryx*. Rather than questioning the, by now, dubious claim that this ancient bird provides evidence for a transitional link between reptiles and birds, many textbooks give no clue that there is any controversy over its ancestry, nor do they point out that modern birds are almost certainly *not* descended from it.

'Wells also considers Darwin's famous finch beak variations. Rather than explaining that selection on finch beaks oscillates between wet years and dry years without resulting in any net evolutionary gain, many books mention these minor, **backwards and forwards** (reversible) variations as if they are evidence for Darwin's hypothesis of long-term, large-scale evolution.

'Wells goes into other Darwinian icons such as peppered moths, four winged fruit flies, the evidence from horse fossils, and **theories about human origins**. The last of which are "controversial, and rest on little evidence." All the famous **monkey-to-man drawings** of ape-like human ancestors are, by the way, hypothetical. Imaginative, but of no scientific validity.

'Casey Luskin adds that, while some people claim that the evidence for humans having evolved from ape-like ancestors is overwhelming, in fact,

> 'hominin fossils generally fall into one of two categories — ape-like species or human-like species (of the genus *Homo*) – and there is a large UNBRIDGED GAP between them. ... the fragmented hominin **fossil record *does not* document the evolution of humans from ape-like precursors**. In fact, scientists are quite sharply divided over who or what our human ancestors even were.
>
> 'Newly discovered fossils are often initially presented to the public with great enthusiasm and fanfare, but **once cooler heads prevail**, their status as human evolutionary ancestors is invariably called into question.
>
> 'The details of the earliest stages of human origins are murky. They come from what UC–Berkeley paleoanthropologist Tim White once called *"a black hole* in the fossil record."[166] ...
>
> 'While early hominin fossils are controversial, due to their fragmented condition, there is one major group – the *australopithecines* – that is *widely promoted* as directly ancestral to humans. The primary claim is that australopithecines had the head of a chimpanzee but a body that allowed it to walk upright, like humans.
>
> '**Despite the prevalence** of that standard view, authorities have found that the fingers, arms, chest, hand bones, striding gait, shoulders, abdomen, inner-ear canals, developmental patterns, toes, and teeth of australopithecines **point away from their being human ancestors** and / or suggest that they didn't have human-like bipedal locomotion.[167] [This evidence includes the world famous australopithecine referred to as 'Lucy.'] ...

'Paleoanthropologist Leslie Aiello, who served as head of the anthropology department at University College London, stated that when it comes to locomotion, "**Australopithecines are like apes, and the *Homo* group are like humans**. Something major occurred when *Homo* evolved, and it wasn't just in the brain."'[168]

'On the Big Bang-like appearance of humanity,

'When the human-like members of our genus *Homo* appear [in the fossil records], they do so **abruptly**. A paper in the Journal of Molecular Biology and Evolution called the appearance of *Homo sapiens* a *"genetic revolution"* in which "no australopithecine species is obviously transitional."[169] In a 2004 book, the famed evolutionary biologist Ernst Mayr explained that "the earliest fossils of *Homo*, *Homo rudolfensis* and *Homo erectus* are **separated from** *Australopithecus* by a large, unbridged G A P" with no "fossils that can serve as missing links."[170]

'... the fossil record provides us with ape-like australopithecines ("before") and human-like *Homo* ("after"), **but not with fossils** documenting a transition between them. In the absence of intermediaries, we're left with "inferences" of a transition based strictly upon the [PRIOR] ASSUMPTION of Darwinian evolution.

'No wonder one commentator argued that IF WE TAKE THE FOSSIL RECORD AT FACE VALUE, it implies a "**big bang theory**" of the appearance of our genus *Homo*. [171]

'... the evidence shows that human-like forms **appear abruptly** in the fossil record, ***without* any fossils connecting us to our alleged ape-like evolutionary ancestors**.

'This contradicts the expectations of neo-Darwinian evolution [common descent] and suggests that unguided [non-intelligent, non-spiritual, blind-chance] evolutionary mechanisms do not account for the origin of our species.'[172]

Alisha said, 'The comment: *"if we take the fossil evidence at face value,"* "it implies a big bang theory of the appearance of humanity," is interesting.

After all, why would we *not* take the evidence at face value? Unless, we have decided that a prior theory, like Darwin's, is the only one possible?'

'I agree, Ali. As so often happens with ***Darwin*** in particular, and with **materialism** in general, too many beliefs and assumptions go unchecked. For example, Darwin and his supporters have endlessly felt that his basic theory is too 'good' not to be true.'

'Which is ideology, Ollie, it's not science…'

'Of course. This is the trouble with Darwinism and materialism today. They are no longer treated as hypotheses open to being tested and falsified but as revealed truths, from St. Darwin, which cannot be questioned.

'There is, in any case, far more that separates people from apes than differences in gait, shoulders, chest shape, inner-ear canals, teeth, and developmental patterns. Take our capacities for abstract thought, and for language and speech. In his book, ***The Kingdom of Speech***, Tom Wolfe notes that not only is the human capacity to speak unique, but that, despite a vast amount of work, over decades, "by some of the greatest minds in academia," all evolutionary theories about the **origins of human language** are incredible, and they fail to explain this unique feature of what it is to be human.

'Michael Egnor writes,

> '… Darwinists are at an utter loss to explain how language… "evolved." … [No] evolutionary ***just-so stories*** come anywhere close to explaining how man might have acquired the ***astonishing*** [and, *biologically,* **totally unnecessary**] ability to craft unlimited [abstract] propositions and concepts and **subtleties within subtleties** using a system of grammar and abstract designators (i.e. words) …
>
> 'Darwin and his progeny have had no dearth of fanciful guesses – birdsongs (Darwin's favorite theory) and grunts and grimaces that mutate… into Cicero and Shakespeare. Evolutionary theorizing about language has been a *colossal waste of time.*
>
> 'None of this evolutionary fancifulness makes any sense, nor has **any real scientific basis**… The human mind has *immaterial* abilities – the intellect's ability to **grasp *abstract* universal concepts** divorced from any particular thing—and that this ability makes us more different from apes than apes are from viruses. …

'We are not just animals who talk. ... Language is the tool by which we think `abstractly`. ... (1) Language is a rock against which [Darwinian] evolutionary theory wrecks, one of the many rocks – (2) the uncooperative fossil record, (3) the jumbled molecular evolutionary tree, (4) irreducible complexity, (5) intricate intracellular design, (6) the genetic code, (7) the collapsing myth of junk DNA, (8) the `immaterial human mind` – that comprise the shoal that is sinking Darwin's [steam age] Victorian fable.' [173]

'Towards the end of *Icons of Evolution,* in a chapter headed *Science or Myth?,* Wells quotes Ernst Mayr, who having once written that,

'it is a considerable strain on one's credulity to assume that . . . systems such as ... the eye of vertebrates, or the bird's feather... could be improved [let alone created—in the first place] by random mutations.'[174]

proposed, in *Scientific American,* that no educated person doubts any longer the 'so-called theory of evolution, which we now know to be **a simple fact.**'

'Wells writes: "Ask any educated person ***how*** we know that evolution is ***a simple fact*** . . . and chances are that person will list some or all of the [dodgy] icons described in this book. For most people— including most biologists—the alleged **icons *are* the very 'evidence'** for Darwinian evolution which they rely on—just as Darwin relied on Haeckel's faked embryo drawings."

'As we have seen, the so called icons of evolution *misrepresent* the evidence. "One icon (Miller–Urey) gives the false impression that scientists have demonstrated an important first step in the origin of life.

"One (the four–winged fruit fly [a deformed and functionally useless byproduct of lab experiments]) is **portrayed as though** it were raw materials for evolution, but is . . . an evolutionary *dead end*. Three icons (vertebrate limbs, *Archaeopteryx,* and Darwin's finches) show actual evidence but are typically **used to conceal** fundamental problems in its interpretation.

"Three (the tree of life, fossil horses and human origins) are incarnations of **[materialist] concepts masquerading** as neutral descriptions of nature. And two icons (Haeckel's embryos, and peppered moths on tree trunks) are *fakes.*

"People such as Ernst Mayer insist that there is overwhelming evidence for Darwin's theory. . . . [Yet,] If there is such **overwhelming evidence** for Darwinian evolution, why do our biology textbooks, science magazines and television nature documentaries keep recycling the **same tired old myths?'**

"There is a pattern here, and it demands an explanation. Instead of *continually testing* their theory against the evidence, as scientists are supposed to do, some Darwinists **consistently ignore, explain away, or misrepresent** the biological facts in order to promote their theory."[175]

'Just like Darwin, Ollie, who was so enchanted by his own theory, overall, that he constantly tended to wave away its many flaws.'

'Yes. With Darwin, it is **always the data that's wrong,** not the theory. The fossils are wrong. The living cell, or blood, or eye, *'might'* have evolved by chance, mightn't it? **Please** suspend your **disbelief**... "Can we not *imagine Cells!, or Blood!, or Hearts!, or Brains!,* evolving in tiny little steps?"

'Well we can imagine anything, can't we, Ollie?'

'Yes, of course. But Darwin and so many of his followers since, cannot seem to tell the difference between an interesting hypothesis, like his, and whether or not the idea, is, in the end, supported by the real data. Darwin's theory, they feel, is just 'too good' not to be right!'

'A case of *"Why spoil a good story with the truth?"* Wondered Ali.

'It feels like it. But it's not how science is *supposed* to be. Wells agrees. He writes, "One isolated example of such behavior might be due simply to overzealousness. Maybe even two. But ten? Year after year?"'[176]

'In his review of *Icons of Evolution,* Dean Kenyon, professor of biology, reflects that Wells has "exposed the exaggerated claims and [self?] deceptions that have persisted in standard textbook discussions of biological origins for many decades, in spite of [all the] contrary evidence!"

'As Milton, Denton, Johnson, Wells, Latham, Behe, Dembski, Meyer, Bethell, Shedinger and many others have shown, it is long past time to let Darwin go.

'In ***Darwin's Bluff*** Robert Shedinger tells how Darwin hoped he might have discovered a ***universal principle,*** a new natural law—almost like gravity—that of a universal, shared, branching, ***tree-like descent***, with nearly infinite changes, of all plants and animals, from just one or two ancestor microbes.

'It was attractively simple, at first glance logical, and, to Darwin's mind, it got around the problem of 'god' or 'spirit' needing to create anything.'

'Although, Ollie, according to Professor James Tour,[177] one of the most brilliant chemists of our time, Darwin was smart enough, in his formal work, to steer well clear of the issue of how *cellular Life* had really begun—the ***actual*** *Origin of Species*—wasn't he?'

'Yes, but his words to his friend, Joseph Hooker, 'But if *(and oh **what a big if**)* we could conceive in some warm little pond with all sorts of ammonia and phosphoric salts… that a protein compound was chemically formed, ready to undergo still more complex changes'[178] [and, by pure chance, become Life], tell us well enough of his anti-supernaturalist views.

'Of course, all this is only a problem for those, like Darwin, with overly simplistic ideas of the supernatural, of the ultimate, *'Spirit-Life-Source-Force-Energy.'* They reject *'Spirit,'* because if we think of *God or Spirit or Intelligent Creative Beingness,* [or other words], as a Father Christmas type figure, making and sustaining everything, *right now!, this amazing second, and every amazing second!, 24/7,* until the end of time, then, of course, most of us will question that.

'But if we conceive of *Spirit-God-Supreme-Intelligent-Beingness* as a *living,* **universal intelligence** which *pervades* and in.FORMS everything, from *Atoms to Cells to Plants, to People to Planets to Stars,* we will have no logical or intellectual need to go for the totally materialistic and anti-supernaturalist kinds of explanations Darwin, and so many of his fans since, have been looking for.'

We finished our snacks, and, resting under the shade of a beautiful tree, fell into a gentle sleep, lulled by the steady murmur of the stream.

16 Can Chance Build Software?

'Darwin, not knowing what we now know, wondered whether an organ as complex and wonderful as an eye or brain could arise *by chance.* He wrote,

> 'IF it could be demonstrated that any complex organ [liver, brain, kidney, womb, eye] existed, which could not possibly have been formed by numerous, successive, slight modifications, **my theory would absolutely break down**. But I can find out no such case.' [179]

'This was odd. Because it should be obvious, to any but the most blinkered, that, not one, but, *many* key features of Living things could not possibly be formed, by 'numerous, successive' lucky-chance 'modifications.'

'Darwin provided no data at all, let alone any *extraordinary* data, to back up his own quite **extraordinary beliefs** in the creative powers of random causation—nor has anyone ever since. As biologist Michael Denton writes,

> 'To common sense [if it is allowed] it seems incredible to attribute such ends [as *Cells, Blood, Hearts, Livers, Eyes, Brains*] to **random search mechanisms, known by experience to be incapable**, at least in finite time, of achieving even the simplest of ends.' [180]

'Denton notes that Darwin himself had tremendous doubts, at times, over the extreme nature of his scientifically untethered assertions. He wrote:

> 'Although the belief that an organ so perfect as the eye could have been formed by natural selection [with no aims in mind], is **enough to stagger anyone** … [*yes!, yes!, yes!.*] I have felt the difficulty far too keenly to be surprised at others hesitating to extend the principle of natural selection to so startling a length.'[181]

'Yet, despite the always highly speculative and utterly unproven nature of his claims, and his own nagging doubts, he was, overall, undeterred. He liked his hypothesis of natural selection too much to be able to simply to let it go, even as it became ever clearer that the real world data did not back him up. In 1885 the Duke of Argyll retold a conversation he had had with Darwin.

"[I]n the course of that conversation I said to Mr. Darwin, with reference to some of his own remarkable works on the *Fertilisation of Orchids,* and upon *The Earthworms,* and various other observations he made of the **wonderful contrivances** for certain purposes in nature—I said it was *impossible* to look at these without seeing that they were the effect and the expression of *Mind.*

"**I shall never forget** Mr. Darwin's answer. He looked at me very hard and said, "Well, that often comes over me with overwhelming force; but at other times," and he shook his head vaguely, adding, "it seems to go away."' (Argyll 1885, 244).[182]

'One scientific way, to have tried to answer his doubts, would have been to provide **proper, probability estimates,** to show that the paths to such complex mechanisms—as *Cells, Blood, Taste, Smell, Sonar!, Kidneys, Eyes, Ears, Brains!,* the caterpillar-to-butterfly enigma, the mysteries of male and female, of asexual and sexual reproduction, testicles, ovaries, wombs, livers, hearts, lungs, and countless other biological systems—could have been found, by **pure, chance,** in the time available. Yet, as Michael Denton writes,

"…**nowhere** in the *Origin [of Species]* is **any attempt** made to provide [any] **quantitative support** for the grand [Darwinian] claim of the all sufficiency of **[mere] chance** [as a great creative principle].

"It is true that Darwin appealed on many instances in the *Origin* to the enormous periods of time available to the evolutionary search but, as is the case with any other random search procedure, time in itself tells us NOTHING of the PROBABILITY of achieving any sort of goal UNLESS THE COMPLEXITY of the search can be **quantified**."[183]

'Interestingly, at the Wistar conference in 1966,

'Some mathematicians did try to make the calculations. The mathematician D.S. Ulam argued that it was **highly improbable** that the eye could have evolved by the accumulation of small [rare mostly harmful, random, genetic] mutations, because the **number**

of [random] mutations would have to be so large and the time available was not nearly long enough for them to appear.

'Sir Peter Medawar and C.H. Waddington **responded that Ulam was doing his science backwards**[!]; the fact was that the eye *had* evolved[!] and therefore the mathematical difficulties must only be apparent[!]. Ernst Mayer observed that Ulam's calculations were based on assumptions that might be unfounded [**while not questioning his own assumptions!**], and concluded that "somehow or other[!] by *adjusting* these figures[!] we will come out all right[!]. We are comforted by the fact that evolution *has* occurred. [!]" [184] [Yet, this is **the very issue.** Has it occurred? Saying 'eyes' exist, therefore they **must have** 'evolved' is question begging.]

'That's interesting, the mathematicians realized that Darwin's ideas could not account for the eye but, because their results conflicted with Darwinism, it was the mathematicians, not Darwin, who *'had'* to be wrong?'

'Well, strangely, to question Darwin's never proven ideas is risky. If you do so, you are likely to be seen as an academic trouble maker or religious literalist—when neither may be true. Darwin's theory is based on rAndoM causation as its supreme creative principle. Yet, there is **no evidence,** from any time or place, that any **purposeful mechanism** can rise from raNDom **causes**—not even a simple wooden toy. Can you, *honestly,* think of any?

'As to the *within-the-cell* level, biologist, Michael Denton, writes,

> "Molecular biology has also shown that the **basic design** of the cell system is essentially the **same in all living systems**, from **bacteria to mammals**. In all organisms the roles of DNA, mRNA and protein are identical. . . . In terms of their basic biochemical design, therefore **no living system** can be thought of as being **primitive or ancestral with respect to any other system, nor is there the slightest empirical hint of an evolutionary sequence** among all the incredibly diverse cells on earth. For those who hoped that molecular biology might bridge the gap between chemistry and biochemistry, the revelation was profoundly disappointing." [185]

'Remember too, that,

"**it is** MORE LIKELY that you *and your entire extended family* would win the state lottery EVERY WEEK for a MILLION YEARS than for a BACTERIUM [or, I would add, white cells or red cells, or livers, or lungs, or kidneys, or brains, or eyes, or ears, or tongues, or teeth, or hearts, or bones, or ovaries, or legs, or tendons, or intestines, or spines, or wings, or feathers, or lungs, or butterflies, or bombardier beetles, or spiders or spiders silk, or their amazing webs] to form by chance!'[186]

'Yet, Ollie, it is hard to re-open people's minds on all this, isn't it?'

'Yes, because most of the big voices in science dismiss the *spiritual* out of hand. Never noticing *how much arrant nonsense* they have to convince themselves to believe for their scheme to *seem* true. They entrance themselves into believing in **pure chance** as a supreme creative force, even though this is not merely illogical, but *demonstrably* impossible. Yet, point this out, and *we* will be the cranks. We live in *strange, upside-down, times...*

'Yet, despite these troublingly unscientific views, Ali, some of the more agile intellectuals of our time *are,* like *Sleeping Beauty,* beginning to awake, and to wonder whether Darwin's paradigm of *unintelligent* design actually works? For example, Antony Flew, who after decades of a very public atheism, decided to FOLLOW THE EVIDENCE as described in, *'There is a God: How the World's Most Notorious Atheist Changed His Mind.'* [187]

'What caused him to have a rethink?' Wondered Alisha.

'For one, Life's **DNA CODES**, which, he realized, could never arise by chance. **CODES** are ABSTRACTIONS, they are not **'material'** are they? They express I D E A S and I N S T R U C T I O N S. The dna **ink,** made of the amino acids, A.C.T. & G., is **not** the `abstract` CODE.

'So it's as though a computer language, like JAVA, was called Silicon Code or SC after the physical materials it is conveyed by?'

'Exactly. This is why some people have suggested, "why not let DNA stand for "Designed Natural Algorithms."[188] Because CODES always consist in meaningful information which can be conveyed in many forms. Even world famous atheist, Richard Dawkins, acknowledges this. He writes,

"What has happened is that genetics has become a branch of information technology. It is pure information. IT'S DIGITAL INFORMATION. It's **precisely the kind of information** that can be **translated digit for digit, byte for byte**, into any other kind of information and then translated back again." [189]

'Although, unlike Flew, Dawkins does not seem to notice, this is "precisely the kind of INFORMATION" which only ever has **intelligent** causes, and *can* only ever have intelligent causes. It is not rAnDOm, or meaningless information, which is sometimes referred to as Shannon Information.

'As Stephen Meyer explains in *Signature in the Cell: DNA and the Evidence for Intelligent Design,* exploring the origins of Life's Codes, no amount of randomly tumbling the DNA letters (A, C, T, & G) around, will ever create Life's software: A G G A A T G C + G G C C T A T A + G T A C C C T T... Mere chance cannot generate even a simple sentence like, 'P L E A S E + F E E D + T H E + C A T.' The random mixing of letters can only generate Shannon information—giBberiSh—nothing functional.

'Further, as Michael Polanyi, the distinguished chemist pointed out, *if* the sequences of DNA letters in life's codes were *physically* determined, by nothing more than the REGULAR LAWS of chemistry,

> "then such a DNA molecule would have no [meaningful] information content ... [Because, to be of any use, the sequence of DNA letters] must be as physically *indeterminate* [able to be arranged in any REGULAR or *irregular* sequence] as the sequence of words is on a printed page." [190] [Regular = ABC + ABC + ABC Irregular = A G G A A T G C + G G C C T A T A.]

'Ironically, far from being unintelligently designed, so sophisticated *are* LIFE'S CODES, that there is, now, research into the use of DNA for data storage and computing. *Futurism.com* announced: "Self-Replicating 'DNA COMPUTERS' Are Set to Change Everything." ... researchers at the University of Manchester are "working on turning strands of DNA into the next basis for computing." Because DNA'S double helix can take two paths simultaneously, it can solve problems quicker.' [191]

'So why don't more people accept all this, like Flew, and move on?' Ali looked sad. She said, 'I don't understand *thiS **dark** dreary wish* to suck the *Life* and the *Soul,* the magic and the meaning out of everything—*to de-intelligence, de-purpose, and de-Soul* Nature?'

'I don't know either, Ali. You'd think scientists would be fascinated to reawaken to the ancient insight that intelligent Life has intelligent sources.'

'Could it be because conceding that the *super* natural is real might lead to a loss of status for today's 'high priests,' the scientists?' Wondered Alisha.

'Well, it is true that, in the 19th century, in a dramatic intellectual reversal, it was the scientists who replaced the philosophers and theologians of the western world as the arbiters of the most important truths. It would not be surprising, then, if their *intellectual* descendants did not wish to give up that glittering prize. We all have egos. Observe how people now hang onto every pronouncement of famous modern scientists like Richard Dawkins, or the late Stephen Hawking, and how much less notice they take of what religious officials may have to say. It is a tremendous role reversal.

'On the other hand, it is becoming ever clearer that the idea that Life is just an accident which evolved by accident doesn't work. Thomas Nagel's book, **Mind and Cosmos: Why the Materialist Neo-Darwinian Conception of Nature is Almost Certainly False** reflects this same, gathering trend.

'He writes that a science so "dependent on SPECULATIVE Darwinian explanations of practically everything,"[192] and so opposed to the classical *scientific (not religious)* hypothesis of *intelligent causation,* is at a dead end. Such an approach cannot provide us with a full explanation of reality, including, especially, $CONSCIOUSNESS$ and $MIND$. Nagel continues,

> "in light of *how little* we *really* understand about the world it would be an advance if the scientific establishment could liberate itself from the materialism and **Darwinism of the gaps** – to adapt one of its own pejorative tags."[193]

'He adds that such a science is *"incapable* of providing [us with] an adequate account, either constitutive or historical, of our universe."

'He dismisses the materialist view of Nature which "purports to capture life and mind through its neo-Darwinian extension" pointing out that it is, "**antecedently unbelievable** – a heroic triumph of **ideological theory** over commonsense..." adding that he "would be willing to bet that the present right-thinking consensus will come to **seem laughable** in a generation or two."

'This is why we, Alisha, unlike Darwin and his friends, who knew so much less than we, can **safely conclude,** as a matter of fact, not theory, that there is nothing in the laws of chemistry, unguided, to produce living things.

'This is why the "aim of the modern movement in biology . . . to explain all biology in terms of **physics and chemistry**,"[194] has not worked—and will *never* work. This is why, Stephen Meyer points out that,

> 'During the last forty years, molecular biology has revealed a **complexity and intricacy of design** that **exceeds anything** that was imaginable during the late-nineteenth century [steam age]. We now know that organisms display any number of distinctive features of **intelligently engineered high-tech systems: information storage and transfer capability**; functioning codes; sorting and delivery systems; **regulatory** and **feed-back loops**; **signal transduction circuitry** ... the [overly] **materialistic science** we have inherited from the late **nineteenth century**, with its exclusive conceptual reliance on matter and energy [cannot] account for the biology of the information age.' [195] [Darwin's theory of 'natural selection' cannot begin to explain this level of complexity. It is not due to good quality science, but due to the *anti-spiritual* ideology of **materialism** that we are still saddled with Darwin's steam age ideas.]

'Stephen Meyer remarks that students of natural history, like Darwin, who cannot do lab experiments to recreate the past, must choose between **competing hypotheses** when trying to make sense of Life's history. The ones they choose should be most in line with their **real world experience** of causes and effects. For example, purely PHYSICAL causes, like the wind and the rain, are, he says, '**causally adequate**' to produce random effects, like landscapes, but *not* CODES or Life Forms.

'He continues that our knowledge of "causal adequacy" provides us with **experience-based** criteria "by which to test, accept, reject, or prefer competing historical theories" [196] which seek to explain past events.

'This is why he says that when a theory, like intelligent design (ID), cites causes, like *intelligence,* which are *known* to produce the effects in question —complex mechanisms—they meet the tests of (i) causal adequacy (ii) of being realistic. This is why, Darwin's ideas—based on the laws of ***unintelligent*** design, of pure chance—fail the key tests of (i) causal adequacy and (ii) of being realistic. Meyer writes,

> "because experience shows [us] that an intelligent agent [or agents] is ... **the only known cause of specified, digitally encoded information,** the theory of intelligent design ... has passed" the tests of causal adequacy and of having the best and most logical explanatory power. This is why **intelligent** design—not mere chance or *unintelligent* design—"stands as the best explanation of the DNA [CODING] enigma." [197]

'You know something, Ali, we know perfectly well, unless we would rather not know, that only *intelligence (mind)* accompanied by the *spiritual* qualities we know as *desire, imagination, creativity*—can create clever mechanisms be they *PCs, TVs, Cells, Hearts, Eyes, Brains,* or **DNA Codes**, not regular laws unguided, not mere chance, nor some mix of the two.'

We continued on our way.

17 What Is Life, What Does Science Say?

'While we, rightly, respect the achievements of modern science, one of its major disappointments has been that it has no compelling ideas as to what *Life* and *Mind* truly are. All it can offer us is Darwin's theory, which, despite the misleading title of his most famous work, *On the Origin of Species*, does not tell us—*at all*—how the species originated, nor, indeed, *Life* itself.

'Worse, since we abandoned Cartesian dualism, we have been led into yet more confusion when it comes to considering *Life* and its true *Origins*.

'Cartesian dualism?' Asked Alisha.

'It's the classical idea, described by the French philosopher, René Descartes, that we can *think* of the body as a kind of MACHINE, but one which, unlike our machines, is infused with *Life, Soul, and Spirit* aspects.

'According to this view, reality—which, in essence, is **a unity (or non-dual)**—manifests through various two-sided aspects, by analogy, like a coin, *Heads* (one side) + TAILS (the other) conceived, variously, as *Spirit* + MATTER, *Soul* + MECHANISM, *Life* + FORM, *Mind* + BRAIN, *Subjective,* **inner,** *living,* ***Soul****, interfused,* temporarily, with natural, physical, BODY.

'In this dualistic understanding of *unitive* reality, there are *Life* and *Soul* principles, the *subtle forces* which make living things *alive*: the *animate* versus the **inanimate**, the *Soul* versus its temporary, PHYSICAL housing.

'But, the dismissal of dualism leads to MATERIALIST-MONISM where only the physical is real, and all *subjective* phenomena, like *thinking* and *feeling,* are held to be **mere** (accidental) **byproducts** of a matter which is, itself, MINDLESS. Francis Crick summed up this arid way of thinking. He wrote,

> "**You, your joys** and your sorrows, your memories and free will, are in fact **no more than** the behavior of a vast assembly of **nerve cells** and their **associated molecules**. As Lewis Carroll's Alice might have phrased it: "You're nothing but a pack of neurons." [198]

'But why do we *still* take seriously, Ollie, the statements of those who claim that anything *we, or they,* say is nothing more than the *mindless* firing, of *mindless* neurons, governed by *mindless* laws, from a *mindless* source?'

'Yes. *Why* do we? It's a mystery to me. Yet, it is precisely because of our acceptance of this unreasonable ideology, that our sciences cannot, now, think any deeper as to **what really** causes Living things to Breathe, our Hearts to Beat, our *Souls* to Feel, or our A W A R E N E S S E S to be Aware.

'So scientists today have no real idea what makes anyone truly *Alive or Aware?*'

'No, how can they? Because, according to science today, there are no *subtle, Life Force or Soul* aspects to living things.

'You mean no principles like *Soul, or Chi, or Prana* as the Chinese and Indian civilizations know these things?'

'Yes . . .'

'But why dismiss the classical intuition that such forces *animate* living things? Surely anyone can see that when a Plant, Animal or Person dies, some *vital, animating* force is no longer there, only a SHELL is left behind.'

'Well, it's mostly because someone artificially synthesized *urea* in 1828 and because *Souls* can't be seen coming and going in telescopes.'

'Ha, ha . . . but what has *urea* got to do with it?'

'Well, at one time, before our science became quite so materialistic in its thinking, it was felt that the real existence of the *subtle, organizing* **Vital or Life Forces** was proven by the *difference* between the *organic* or *'living'* chemicals making up the plants and animals, and the INORGANIC OR 'DEAD' CHEMICALS which make up rocks and stones.

'So when, in 1828, someone artificially synthesized *urea,* an organic chemical, from INORGANIC CHEMICALS, the hypothesis collapsed.'

'That's quite a jump, isn't it?' Said Alisha.

'Yes, because it's not just the differences in materials which tell us whether something is intelligent or random in origin—but its *functionality.*

'We don't say, "No intelligence goes into a TV," just because the 'organic' (organized) wires and plastics in it are, in essence, no different from the 'INORGANIC' (unorganized) ores and oils in the ground from which the wires and plastics in the TV were derived, by *intelligent* forces using *intelligent* means—albeit ones far less sophisticated than those required to organize even the most basic Life form, let alone cell, blood, heart, lung or wing, all far more complicated than anything we make.[199]

'Yet, if we think it through, according to Crick's view, the current *'scientific'* view, we are no more truly *'alive'* than the rocks and the stones.

'Anima, after all, means *Life and Soul,* but the current 'science' stance is that we do not have *Life Forces,* nor even, real *Minds,* let alone *Souls.'*

'Why not?'

'Because, although the urea didn't synthesize itself—a human *intelligence* did—the fact that it could be artificially synthesized led some to decide that this falsified *vitalism.* So, since then, to be a vitalist, or to believe in the *Soul,* in the *super-natural,* or *intelligent design,* is held to be disreputable—especially in the academic west.

'This is why the hoard of modern evidence for out of body experiences (OBEs), near death experiences (NDEs) and for *after-life* communications, from those passed on, even when it is verifiable,[200] is not taken seriously. This has been the attitude, for well over 100 years now.

'Yet, the honest truth, Ali, is that no one has any idea, scientifically speaking, how *Life* or *Mind* arise, in the remote past or, *right now!, 24/7, this amazing second . . . and every! . . . amazing . . . living! . . . second! . . .*

'There are, literally, **trillions!** of biochemical operations happening in our bodies, *right now!, this second, and every second!, 24/7,* to keep the whole, astonishing thing working, but it is all due to **pure chance?**'

'Ah, Ollie, we think of ourselves as a rational culture, yet some of the S U P E R S T I T I O N S *we* now hold—in the name of sober science—are worse than the old ones! At least the religions, for all their faults, always understood that there is a basic *intelligence* and *meaning* to reality.'

'I agree. Darwin's idea that microbes might arise by chance, then, morph, via many stages, into Lions with their mysterious Cat consciousnesses, Wolves with their intriguing Dog natures, full of potential love and friendship for people, and, finally, into Humans with all their extraordinary gifts, has **no scientific basis**. It was just a strange new *faith,* even if, for Darwin to wonder whether Life might just be an amazing accident, which evolved by accident, was just about understandable . . .'

'Because he and his steam age colleagues knew so much less than we do of the truly mind-boggling complexities of living things?'

'Yes. But for *us* to continue to "believe the impossible," as George Wald suggested we do, "on philosophical grounds," no longer makes sense.

'I remember Richard Dawkins get quite cross once because someone suggested to him that he, Dawkins, believed in mere CHEMICAL ACCIDENT as the origin of living things. He replied that he believed no such thing.'

'But if, like Wald, or Crick, or Dawkins, you rule out *super* natural causation, either as a matter of your scientific method (MM/MN) or your philosophy (PM/PN),[201] as they do,' mused Alisha, 'what are you left with? Pure 'luck' HAS to be your 'god,' it HAS to be your great creative principle.'

'Yes. Even if, by his comments, Dawkins meant that the regular laws of chemistry, unguided, could lead to Life, he is obliged by his atheistic philosophy to believe that those laws are, themselves, mindless in origin.'

'Which is the same thing as saying that Life arose by chance?'

'Yes. While some dismiss all the evidence falsifying Darwin, no one, today, who is paying attention, has any respectable scientific excuse ***not*** to concede that nature's REGULAR LAWS (amazing as they are) + *iRreGulAr* **chance** (mysterious as it is) will ***never*** lead to living things, even of the most basic kind.

'Recall, "**it is** MORE LIKELY that you *and your entire extended family* would win the state lottery EVERY WEEK for a MILLION YEARS than for a bacterium to form by chance!'[202]

'We need to remember, Ali, that science does not = MATERIALISM and MATERIALISM does not = 'rationality.' A true rationalist is one who considers the data and applies their logic, reason, and intuition to it at every stage. *Life and Mind* are deep mysteries. No one, who thinks, can deny it.

'But those who assert that, "Only the PHYSICAL can be real," are no more scientific than the spiritual person who says, "*Spiritual Forces* do it all, so let's not bother trying to work out *how or why* they do so." Neither is scientific. Because the first is question begging, and it needlessly restricts scientific inquiry, while the second, is lazy and incurious.'

18 Mind Before Matter: Intelligent Design

'How many of us, Ali, ever truly contemplate our *thinking* and *feeling?* As far as we know, the PHYSICAL ATOMS of which our bodies are made, have no capacity to *think* and to *feel,* to see and to hear. Where, then, is the *'science'* in the assertion that our amazing capacities to *think,* to *know,* to *feel* and to *love,* are mere *accidents* of a MATTER which is itself *mindless?'*

'It is strange, Ollie, that so many of those, today, who consider themselves scientific and 'rational' really do not seem to understand that no amount of examining BRAIN CHEMICALS will ever locate our *immaterial* thoughts, or our immaterial ideas, anymore than analyzing the PARTS from which a TV (the BODY) is made will ever find the *Show (Soul)!'*

'Yes. As the distinguished philosopher, Thomas Nagel writes,

> "The existence of consciousness is... one of the most astounding things . . . if physical science... leaves us... in the dark about ***consciousness,*** that shows that **it cannot provide** [us] the basic form of intelligibility for this world." He continues,
>
> "For a long time I have found the **materialist** account of how we and our fellow organisms came to exist **hard to believe**, including the standard [Darwinian] version of how the evolutionary process works. **The more details we learn about** the chemical basis of life and the intricacy of the genetic code, **the more unbelievable the standard historical account [Darwinism]** becomes...it seems to me that. . . the **current orthodoxy** about the cosmic order is the **product of governing assumptions that are unsupported**, and that it FLIES IN THE FACE OF COMMON SENSE.' [203] [Ah yes!]

'That's refreshing, an intellectual who is actually *thinking!'*

'Yes, although, he's still quite rare. He's right, though. The materialist account of life *is* hard to believe. Even if it isn't possible, for most of us, to see *Soul* and *Spirit* directly, *really,* we can *see* and *hear* them all around us. They are hardly hiding. We can see them in the stunning complexity of cells, in the butterfly, in the peacocks tail, and the eagle's magnificent wing.

'We can hear them in our friends laughter and our children and pets warm greetings. We can, even, just quietly look in the mirror.

'In, *Irreducible Mind,* the authors note that, the current mainstream view in "psychology, neuroscience, and philosophy of mind holds that all aspects of human M I N D and C O N S C I O U S N E S S are generated by [purely] PHYSICAL processes occurring in [purely PHYSICAL] brains. ... **this reductive materialism is not only incomplete but false."**

'Why? Because, too many phenomena are **"impossible, to account for"** in PURELY PHYSICAL terms, "memory…near-death experiences and allied phenomena, genius-level creativity, and 'mystical' states of consciousness" [204]

'Raymond Tallis, a neuroscientist also criticizes the materialist urge to reduce people to nothing more than their BODY-BRAINS. In *Aping Mankind: Neuromania, Darwinitis and the Misrepresentation of Humanity,* he blasts,

> 'the *exaggerated claims* made for the ability of neuroscience and evolutionary theory to explain human **consciousness,** behavior, culture and society [which he calls 'neuromania']. Tallis dismantles the idea that **"we are our brains,"** criticizing a "plethora of neuro prefixed pseudo-disciplines," that sidestep "a whole range of mind-body problems. . . . offering a **grotesquely [over] simplified and degrading account of humanity.'**[205]

'In *Science and the Near Death Experience,* Chris Carter comments that materialists believe that "C O N S C I O U S N E S S is created by MATTER, yet, he says, the *best theory* we have about the nature of matter [quantum theory] seems to require that C O N S C I O U S N E S S exists **independently** of matter.' [206]

'The challenge for materialists is that if C O N S C I O U S N E S S can exist apart from matter, their entire premise is shattered. Because it reestablishes humanity's ancient understanding that C O N S C I O U S N E S S and INTELLIGENT MIND are as much fundamental properties of reality as the 'mindless matter' which materialists worship. As John Searle writes,

'Acceptance of the current [materialist] views is motivated not so much by an independent conviction of the truth, as by **a terror** of ... [the] alternatives. That is, the choice we are tacitly presented with is between a 'scientific' approach, as represented by ... '**materialism**,' and an 'anti-scientific' approach, as represented by... some... traditional *religious* conception of the mind.'[207]

'Unfortunately, so many of us now conflate science with materialism, as though to be rational, one must be a materialist. This is why some scientists promote the idea that *spiritual mind* could not influence 'non-spiritual' MATTER, because, they argue, how could *'unlike,' immaterial Mind*, act on 'LIKE' (MATTER)? How could *Mind* (the *'Signal'*) exist apart from the **body-brain** (the physical 'TV'), nor, if apart, how can it *influence* it in any way?

'Which, if true, Ollie, would mean that an *intelligent* evolution, driven by *spiritual intelligence* would not be possible? There would be no way *in* for *Creative Spiritual or Divine Mind* to influence the **material** worlds.'

'Yes. That's their stance. But there are many ways to falsify the claim that *Mind* cannot exist apart from, nor that it can influence MATTER.

'There is, for example, much modern evidence that **mind-to-matter** communication *is* **possible**. It has been repeatedly shown that people *can remotely* influence random number generators.[208] Many have also described near death[209] and out of body[210] experiences, pointing to *non-physical* dimensions, and to the existence of *mind* beyond the PHYSICAL BODY.

'As Darwin's colleague, Alfred Russell Wallace, saw, gifted psychics have demonstrated, in many carefully controlled experiments,[211] that they can transmit *verifiable* messages from those passed on. The *Soul* survives 'death.'

Alisha looked thoughtful. 'Hmm,' she said, 'so, if human *minds* can influence random number generators, if we, as *consciousnesses,* can operate *beyond* the body, and if, *after* death*,* it is possible to communicate with those still in the physical, it implies that there may be *other levels of Mind*, of *multi-dimensional Mind,* all the way up to *spiritual or divine mind?'*

'Yes. If our minds are *multi-dimensional,* then *higher* levels of mind than ours, including *spiritual* mind, are also likely real—as the psychics and shamans, the yogis and the sufis have always said. This means that an

intelligent creation or evolution, ruled by consciousnesses far higher than ours, is also possible. **Intelligent** selection *vs.* MINDLESS selection.

'Neal Grossman, a philosopher of science, was once asked *how* could *Mind,* the mysterious source, in us, of *thinking and feeling,* if it, truly, is immaterial, act on the BODY, the PHYSICAL MECHANISM, the 'TV'?'

'Why? Because some scientists theorize that our universe is a closed system. So, they ask, "If our theory is correct, how could there be **any way** *in* for *spiritual* Mind to *create* or to *influence* the physical worlds?"

'Neil Grossman addresses these concerns. He says,

> 'On general principles, it is unscientific to claim that one's current theory is true a priori. Facts trump theory. Psychokinesis [mind-to-matter interaction] **is a fact** that has been unequivocally demonstrated over and over again (e.g. the PEARS experiments).
>
> It is absurd to deny that the facts are what they are simply because their occurrence poses difficulties for our current theories.
>
> It is ***not*** **good scientific practice** to reject empirical data because it poses problems for conventional theories. Materialism is clearly *false* ... The point is that we must stay close to the data, even if there is [currently] no theory that adequately explains it.
>
> 'Anomalous data is the engine that drives the search for new theories, even though, as the history of science shows, there will always be those who **prefer their old theories** to new data that falsifies what they believe.
>
> 'So, **it is a fact** that mind can influence a random number generator; we do not now have a theory that can explain this fact, and it is highly unlikely that such a theory will be compatible with Materialism.'[212]

'Grossman is right. "Facts trump theory," or, at least, they should, even if this ***does not*** always happen, as we see repeatedly with Darwin and his ideas. The theory has been *falsified,* many times in fact, but we take no notice. Just as in Galileo's day, when we get too attached to a set of ideas—be it geocentricism, materialism or *creation/evolution* by blind-chance-ism—it can be difficult to convince us that other ideas may be better.

'In *Science and the Near Death Experience,* Chris Carter notes that many people now think that cartesian dualism—*Mind vs* MATTER, *Soul vs* MECHANISM, *Life vs* FORM thinking—is outdated. Not all scientists think this way. Carter quotes the astronomer V. A. Firsoff, who wrote,

> 'to assert there is *only* **matter and no mind** is the most illogical of propositions, quite apart from the findings of modern physics, which show that **there** *is* **no m....a....t....t....e....r** in the traditional meaning of the term.'[213] [There is just energy-space...]

'Carter notes the irony that, while biologists have been trying to bring biology into line with the very MECHANICAL and very reductionistic thinking, typical of 19th-century physics, where, as Francis Crick put it,

> 'The ultimate aim of the modern movement in biology is to [try to] explain all biology in terms of [lifeless] physics and [mindless] chemistry,'[214] [Which is a good definition of scientific reductionism.]

'While, as Harold Morowitz, professor of molecular biophysics writes,

> 'physicists, faced with compelling experimental evidence, have [by contrast with their colleagues in biology,] been moving away from strictly MECHANICAL models of the universe to a view that sees the *mind* as playing an integral role in all physical events.' [215]

'Carter adds, "we have known since the early years of the 20th century that **classical physics fails drastically** at the atomic and subatomic levels, and that the behavior of such particles is deterministic and observer dependent. **The irony here is that while materialists often describe themselves as promoting a scientific outlook,** it is possible to be a materialist *ONLY BY IGNORING* the most successful scientific theory of matter the world has yet seen."

'He continues, "The materialist believes that *C O N S C I O U S N E S S* *[the Signal]* is created by matter [the BODY, the TV], yet the **best theory we have** about the nature of matter [quantum physics] seems to require that *C O N S C I O U S N E S S* **exists independently of matter.**' [216]

'Carter also tackles the idea that the universe can '*only* be physical' and, if this is true, then no *spiritual* event can ever cause a PHYSICAL event.[217] But, Carter says, "we know now that the **universe is *not* a closed system** and that the collapse of the wave function – a physical event – is linked with an antecedent *M E N T A L* event. The objection K. R. Rao describes is of course based on classical physics."[218] He quotes Rosenblum and Kuttner, who write,

> "Some theorists deny the possibility of *duality* [mind-to-matter-causation] by arguing that the signal from a *nonmaterial* mind could not carry energy and thus could not influence MATERIAL brain cells. ...
>
> 'However, **no energy** need be involved in determining to which particular situation a wave function collapses. . . . **Quantum mechanics therefore allows an escape from the supposed fatal flaw of dualism**. It is a mistake to think that dualism [mind-to-matter causation] can be ruled out on the basis of physics.' [219]

'Carter adds that Karl Popper noted that the belief that only *like* can act upon *like* rests on **outdated ideas** about physics. Examples of **unlikes** acting on each other are the interactions between the four, very different forces in physics: gravity, electro-magnetism and the strong and weak forces.

'So the *idea* that the universe cannot be influenced by *Mind,* including *Spiritual or Divine mind* is false?'

'Yes. This discussion, although it seems a bit obscure, cuts to the *heart* of all we have been contemplating. Is *spirit* ontologically real or not? Is *mind* a cosmic principle in its own right? Or is it just a cosmic accident?

'Materialists argue that CONSCIOUSNESS and MIND are merely **accidental byproducts** of a MATTER which is, itself, totally MINDLESS and UNINTELLIGENT. This, they want us to agree, is the 'scientific' view, the 'rational' view. This is why the famous, 20th century philosopher, the late Daniel Dennett, made the following, utterly bizarre, statement: "We're all zombies. Nobody is conscious."[220] In other words, consciousness and mind are not essentially real. They may appear to be, but that's a total illusion.

'Really, Ollie? Have we lost our minds? In the name of *science?'*

'Well, sadly, dogmatic materialists, like Dennett, give illogical voice to an *irrational* rationalism, which, in the name of holy logic and sacred reason, conflates 'rationality' with the cult of MATERIALISM, when this is the last thing any of us should do, let alone some of the brightest and most intellectually gifted among us.' We continued on our way.

19 The Trouble With Hume

'Contrary to its detractors, the main claim of modern ID thinking is **not** that living things **cannot evolve,** but that they cannot arise, or evolve, by the wholly mindless means Darwin and other materialists argue for.

'Whether an overarching *Higher Nature,* to which humanity gives its traditional religious names, can only *create, but cannot evolve* things, is a matter for religious debate. It is not a matter for science.

'It is not, though, solely a religious matter to wonder whether there may be *more* to reality than the purely physical. Even Richard Dawkins has argued that the supernatural can be a hypothesis of science. He just feels that it is a hypothesis which has been disproved. We do not need to be religious to disagree. Because, as we have seen, it is possible, today, to *scientifically* detect evidences for intelligence and design in nature.'

'So, if our sciences became more open, once more, Ollie, to wondering whether physical nature, and the laws we know of, so far, may be derived from an overarching *Spiritual Nature,* it would be uncontroversial to say that all things have *'natural'* explanations, wouldn't it? Because such a posture would allow for the more *subtle 'natural'* elements of reality.'

'Yes, it's true. But if we insist that only PHYSICAL nature can be real, with *no intelligence* to *any* of it, including, if we are to be consistent, to ourselves, we are not going to wonder when or where scientists may be looking at *intelligent processes* in nature, versus random.

'But for those who do not have "a prior commitment to materialism," [221] in the way that Lewontin urges, it is not unreasonable to wonder if intelligent causation may be part of reality and to argue for it.'

'So how do you do that?' Asked Alisha.

'Firstly, **logical** arguments. Is it logical to argue that purposeful *mind* is an accident of MINDLESS matter? Isn't that logically self-refuting? Is it **logical** to hold that human beings are mere BIO-ROBOTS, which, unlike all robots in the real world, mere 'coincidences' *built?* Is that honestly ***scientific?*** Just swirl the robot parts about and all will be well? Let's put all our real-world experience of robot building on one side?

'Secondly, **mathematical** arguments. CODES and software do not arise by chance. Why not? Because *they cannot*. Yes, we can ask, as a somewhat pointless thought experiment, "Could Life's DNA software—unlike all other software—arise by chance?" *'No, it's impossible,'* the answer swiftly comes back. Are we to take no notice? Because it falsifies our ideology?'

"Could Cells, the most complex mechanisms known to us, arise by chance?" "Hmm… **no,** the laws of probability, the laws of entropy, the laws of logic, the laws of commonsense, all forbid it."

'Thirdly, **empirical** arguments. We know that accidents in nature produce tsunamis, avalanches, and wild fires. But anything *functional?'*

'Prior to Darwin, intelligent causation was taken for granted. For example, the philosopher William Paley pointed out that anyone who came across a WATCH, as compared to a rAnDom rOck, could easily work out that it was designed, and not some random, hiGgLe*dy* piGgleDy thing.

'He argued that the Life Forms were **non-random** mechanisms with machinelike characteristics which ***could not possibly*** be 'natural accidents' of chemicals any more than a WATCH could possibly be a 'nAturaL aCciDent' of its **specifically ordered** parts. He wrote, "**Every indication of contrivance, every manifestation of design**, which existed in the watch, exists in the works of nature; with the difference, on the side of nature, of being greater or more, and that in a degree which **exceeds all computation.**" [222]

'No one would question Darwin's, once novel, paradigm of ***unintelligent design,*** if the way we made Clocks, Cars, or PCs was to put rocks in containers, randomly tumble them for a while, then take out the finished products. Yet, all the Life Forms are far more complex than ***anything*** we make, using *all* our intelligence. [*But they arose by pure, unintelligent chance? For no reason, or point, or meaning at all? Who seriously believes this?*]

'What happened, then, to William Paley's design arguments?'

'They were abandoned partly because so many so credulously took up Darwin's forever *unproven **speculations,*** those "BRIEF EXTRACTS" from his allegedly much larger manuscripts, which were, as yet, he said, "**imperfect,**" *without* "**references,**" only "**general conclusions,**" with just a "FEW FACTS in illustration." and partly because of the philosopher David Hume's claim that living things are not really comparable to machines and, therefore, according to Hume, Paley's design arguments from analogy were invalid.

'However, while Hume seems not to have noticed that arguments from analogy are *not* intended to compare precisely like with like, if they were, they'd be pointless, it is true that they can only go so far. In this case, though, their limitations ***hinder*** rather than help Hume's argument.'

'Why?' Asked Alisha.

'Because *living things,* while they can be compared to MACHINES in some ways, are *far* more complex than any machines we make. Yet no machine of ours *ever* arises by chance. So how could nature's ones do so?

'Secondly, because Life's DNA CODED instructions are not *'like'* code, which would be an argument from analogy, and subject to its limitations, they ***are*** CODE. Code more sophisticated than any code or software we make.

'Bill Gates has said of Life's DNA software that it is, **'like a computer program but far, far more advanced than any software [we've] ever created,**'[223]

'Is he right?'

'Yes, of course he is.'

'Which contradicts the "reality/nature-is-**not**-intelligent" worldview?'

'Yes, of course it does.'

'But does that prove anyone's specific religious or spiritual claims?'

'No. But it points to intelligent causation *not* to blind chance!

'Another of the disappointing things with our science, now so captured by materialist thinking, is that not only can it be ***self-contradictory,*** it can be tricky. One minute it tells us that Nature is purely MECHANICAL, and all her mechanisms are just *lucky* coincidences. But if you point out that machinery, simple or complex, *never* arises by chance, that 'machinery,' *always* points to ***intelligence****,* no matter its source, known or unknown, your design argument suddenly becomes invalid, because, according to Hume and his followers, you are arguing from analogy, and the Life Forms are, after all, nothing like machinery. This, though, is not merely tricky, it's false.'

'Why?'

'Because the scientists of Darwin's day had no idea just how MACHINE-LIKE many biological mechanisms would, in time, turn out to be.'

'So William Paley was eventually vindicated?'

'Yes, and to a remarkable degree. Eastman and Missler write:

> 'However, the astonishing discoveries in molecular biology during the last 40 years have finally and unequivocally demonstrated that living systems are, in fact, machines – even to the deepest, molecular level! From the tiniest enzyme to the most complex organ systems found in man, Paley's machine analogy is confirmed.
>
> At the enzymatic level we see an eerie resemblance to the design and operation of chemical factories. At the organ level we find **"hardware" of an unimaginable complexity and ingenuity**. In our five senses we find sensory receivers made of multiple components, each machine-like, the operation of which is absolutely necessary for each sense (taste, sight, smell, hearing, touch) to function properly. In the function of the human heart we **see an incredibly efficient and durable hydraulic pump, the likes of which no engineer has imagined**. Finally, in the structure of the **human brain** we find a computer 1000 times faster than a Cray supercomputer **with more connections than all the computers, phone systems and electronic appliances on planet earth!**' [224]

'You know, Ali, the difficulty modern ID thinking poses for Darwin's theory of unintelligent design or UD is not for any lack of intelligence or design in Nature. No, for these are **pervasive and easy for all of us to see**. Nor is it that ID poses any threat to PRACTICAL, EVERYDAY, SCIENCE. Of course it doesn't. The problem is *philosophical* and *political*. ID challenges the modern, materialistic belief system, which, so strangely, urges us to rely on pure stupid chance, to try to explain all things, including, even, our own bright *Minds* and asks us to feel satisfied with this obvious non-sense.

'Hume is also remembered for his skepticism about ***miracles***. What he forgot, is that the *most miraculous* thing of all is that there is *anything-at-all*, rather than nothing at all. Well, what is the *likely nature* of this amazing *All-*

That-Exists-ness, and its *ultimate Source, which just is?* Is it a DUMB 'miracle' or an *intelligent* miracle? Each of us makes our private choice.

> **"In the cerebral cortex alone, there are roughly 125 trillion synapses, which is about how many stars fill 1,500 Milky Way galaxies**. ... One synapse, by itself, is more like a microprocessor ... In fact, one synapse may contain on the order of 1,000 molecular-scale switches.[!] A single human brain has more switches than all the computers and routers and Internet connections on Earth**."** [225]

'Is this good evidence for *'unintelligent design?'* Really? No microprocessor or PC we make, *ever* arises by chance—but nature's designs do?

'If we do *not* join up the dots on this, we are no longer thinking, and we are certainly not being scientific. Yet, since Darwin, so many scientists and other thinkers have put their telescopes to a blind eye, refusing, all the while, to notice that a *stunningly intelligent, spiritual, cosmic order* may exist.

'Perhaps such an admission would be too threatening to some people's world views? To change our minds, *and hearts* especially, can be difficult.'

'I know, Ali. But how can we justify continuing to believe in Darwin's theory of *"unintelligent design,"* just because some people cannot accept the *super* natural—on philosophical grounds? Even openly admitting that they "choose to believe the impossible: that life arose… *by chance!"'* [226]

'Is this why the current *'science'* view is that the ***brain's amazing*** organization arises, at base, from mere chemical coincidence?

'Is this why Darwin's ***unreal paradigm*** of unintelligent design has turned parts of our sciences into hostages to nonsense, and into vehicles, not for enlightenment but, for academic dissembling and downright unreasonableness? For, if we may not call such processes intelligent and purposeful, our language has become meaningless, and important parts of our sciences ways to *distort* our understandings of reality, not explain it to us.'

'Why do you think all this has happened, Ollie?'

'I think it is partly because of the confusing conflation of various ideas, some true, some false, leading to the current mental mess, especially here in the west, where, against all reason, experience, and commonsense, BLIND LUCK has been elevated to a supreme creative principle.'

'Ideas like?'

'(1) Like conflating the scientific method of inquiry and experiment with MATERIALISM, which is a belief system—not 'science.' As Bruce Greyson, a prominent near death researcher, writes,

> "**Materialists** often claim credit for the scientific advances of the past few centuries. But it is the *scientific method* of empirical *hypothesis testing,* rather than a MATERIALISTIC philosophy, that has been responsible for the success of science in explaining the [physical] world[s]. If it comes to a choice between empirical method and the materialistic world view, the true scientist will choose the former." 227

(2) The discovery that Life on earth is far older than people once believed, followed by the *non sequitur* that there can be *no* intelligent causation at all.

(3) Not realizing that *describing* natural history, is *not* the same as Darwin's theory which tries to *explain* natural history, nor is it the same as the idea that only a MATERIALIST explanation of Life, like Darwin's, can make sense of the data, versus the classical hypothesis of *intelligent causation*.

'This muddled thinking means trying to make sense of a confusing mix of propositions, some reasonable and true, others false. For example:

1. The creation story in the Bible is *not* science. *True,* it's a metaphor.
2. So something like Darwin's theory is the only alternative. *False.*
3. Life on Earth is extremely ancient, the fossils prove it. *True.*
4. So *super* natural creation or *intelligent* evolution is not possible. *False.*
5. Life on Earth has changed greatly over great time. *True.*
6. This proves Darwinian evolution is a fact. *False.* Why? Because,

    (a) There is no evidence for *Darwin's core idea* of universal common descent, with huge modifications. Why? Because…

    (b) The *hypothetical* branching points, on his hypothetical tree of life, where the shared ancestors, of later species, *ought,* according to Darwin's theory, to be, are empty.

    (c) No convincing transitional fossils have ever been found.

(4) Darwin's theory **has** to be true because **materialism** is true. *False.* Materialism is just one hypothesis about the basic nature of reality which, as the philosopher Thomas Nagel says, is "almost certainly false."

20 Sorry Wikipedia, no…

'One of the odd things about all this, Alisha, is that despite the fossil and so much other data falsifying Darwin, relatively few scientists and academics seem to care. But, it's certainly not how science is supposed to be.

'But why don't more thinkers notice, Ollie, that the anti-spiritual wish, expressed by Francis Crick, to "explain all biology in terms of *physics and chemistry,*" [228] and nothing more, does not work. *We sit, walk, run, eat, sleep, think, laugh, and cry,* **inside** *our astonishing bodies,* **trillions** of cellular operations taking place **every second!, 24/7!,** and we profess to believe that **all this** *that we are, and* **all this** *which lies, so amazingly,* all around us, arose and continues to arise, due to pure, mindless chance, *PMC?*'

'I could not agree with you more, Ali. But the tilts towards **materialism** and **Darwinism** in our culture are now so great that those who question them, or who propose classical ideas of intelligent design, are, ironically, accused of being *unscientific*. It can even put careers at risk.[229]

'Look up intelligent design in *wikipedia,* for example, and you'll be told, quite wrongly, that ID has been *falsified*. While ID is, currently, [2025], a minority view, it has **not** been falsified, and majority views do not, always, amount to scientific truth. Science is not just about majority views, but about *competing ideas* and the search for truth. ID is a testable hypothesis, just as Darwin's theory of unintelligent design or UD is.

'The classical hypothesis of *intelligent* causation was, until Darwin's time, mainstream. But, currently [at time of writing], the wikipedia entry on intelligent design starts,

> 'Intelligent design is the *pseudoscientific* view that "certain features of the universe and of living things are **best** explained by an *intelligent* [*super* natural] cause, **not** an UNDIRECTED AND AIMLESS PROCESS such as natural selection."'

195

'Hold on Ollie, surely it would make much better sense for wikipedia to say,

"Unintelligent design, UD, ~~is~~ *was* the *pseudoscientific* view that certain features of living things, like **DNA CODES**, *Cells, Hearts, Minds, Eyes,* ~~are~~ *were best explained* by an UNDIRECTED PROCESS. It was Darwin who popularized this deeply unrealistic idea... but, strangely, it took 160 years for it to be dismissed."

'I agree. Yet, the DIY encyclopedia does express the current mainstream *science* view on this topic. But, oh dear, everything is now so weirdly topsy turvy, back-to-front, and upside down. Truth is lies, and lies are 'truth.'

'Theories of *unintelligent* design, like Darwin's, are 'scientific,' but ID is pseudo-scientific! Despite the fact that, before Darwin, all the greatest minds —from Plato and Aristotle to Galileo and Newton, from Leibniz and Boyle to Faraday, Maxwell and Planck—believed in some form of *intelligent design*. Of course they *did!* ID was the classical, pre-Darwin view.

'Wikipedia, which is supposed to be a trustworthy source of information, is misleading us in this area. It does, though, get one thing right: As it says, natural selection is a mindless and "undirected process." This at least is very *true!* But, for getting Life started *or* evolved, it is akin to ***putting chemicals in a barrel,*** swirling it around, then hoping something clever will emerge, again and again and again! This, according to wikipedia, is *'scientific.'*

'Is this why **un-*intelligent* design**—blind chance causation—which is not capable, in the real world, of assembling even the simplest of working things, not even a wooden toy, let alone a Car, PC or TV, all far less complicated than *any* living things, is held up by *wikipedia,* and many otherwise rational institutions, to be the **great creative principle** which led to all Life on Earth? *Honestly?* Wikipedia's silly attack on ID continues,

'Both **irreducible complexity** and **specified complexity** [theories in ID] present **detailed <u>negative assertions</u> that certain features** (biological and informational, respectively) **are too complex** [*yes!*] **to be the result of** [*mindless!*] natural processes. Proponents [like Behe, Myer, Dembski] then conclude, by analogy, that these features are evidence of [intelligence and] design. Detailed scientific examination has rebutted the claims that evolutionary explanations [**mindless laws** + aimless **chance**] are inadequate [to explain *all* natural phenomena].' This is all pure *bluff!*

'*Bluff, bluff, bluff*... the kind of bluff Darwin was so prone to… I would love to ask those who wrote this nonsense: "Can you please show us the "detailed scientific examination" you refer to? Who has shown, ***scientifically,*** that Life really *could,* and really *did,* arise, *and* evolve, by mere chance?"

'No, what wikipedia calls *"evolutionary explanations"* is similar to Darwin's own style of specious, speculative, tendentious claims that:

"Well, this *might* have happened, then that*, and **that**, and. . . voila!,* Living things, the most complex things known to us. Isn't it amazing what **mere luck** can create! Wikipedia, your trusted defender of 'reason' [**materialism**] versus unreason [*Spirit*]!"

'Wikipedia misleads: There is **no evidence** that LIFE'S CODES could write themselves (Meyer, Lennox, Gitt, Flew). It is demonstrably impossible.

'No amount of 'evolutionary' *bluffing,* no amount of "**positive assertions**" that *all* of *Life, Cells!, Eyes, Ears, Hearts!, Lungs, Livers, Brains!, 'might'* be the result of "undirected" or "aimless" natural processes—by which wikipedia means mindless, blind-chance, processes—can get around that.

'Statements like: "Most scientists agree that Life is a purely 'natural' phenomenon"*are **ideology**,* not science. Wikipedia, strangely, spouts nonsense.

'There is **no evidence** that stunningly complex *Living Cells* could arise by chance, (Dembski, Hoyle, Eastman, Missler, Denton). Even some famous materialists (Wald, Monod, Crick) admit it's impossible—before they ignore their *own, logical* conclusions . . . for *philosophical* and *ideological* reasons.

'There is **no evidence** that dinosaur lungs could turn into avian lungs without fatal consequences, 'DEATH!, WITHIN MINUTES . . .' (Denton).

'There is **no evidence** that the amazing metamorphosis of the crawling caterpillar into the beautiful Butterfly could happen by chance, (Wallace).

'To fight all this with misleading statements in *wikipedia* or with 'unsubstantiated *"Just So"* stories,' as Lewontin urges us to do, just because some people insist that "science = MATERIALISM," is a betrayal of the very science which thinkers like Lewontin, and, presumably, wikipedia, so admire and wish to protect. The materialist *belief* system, and the atheism which automatically flows from it, is a set of ideas about ***ontology***, about the ***deep nature*** of things, about ***what really 'Is,'*** which may or may not be true.

'***Science,*** on the other hand, is the claim, well proven by now, that it is possible, using logic, commonsense, the airing and testing of hypotheses, experiments—and, as Einstein said, our imaginations and intuitions—to inquire into the true nature of reality, into the ***true nature of what really Is.***

'Whether *'what Is'* is just physical, or whether it consists, also, *in mental, emotional, and* spiritual *dimensions,* we can attempt to explore all of it.

'Darwin **could not find one real world example,** ***not one!***, to demonstrate the allegedly so very amazing creative powers of *aimless—mindless—unintelligent—CHANCE*—through the *mindless, aimless* coming together of *mindless, 'purposeless' chemicals—for no reason at all*—to create anything ***remotely intelligent*** *at all*—and no one has ever found any since.

'Yet we are asked by wikipedia, and many otherwise rational bodies, to treat ID as pseudoscientific, but theories of ***unintelligent design,*** UD, of blind **chance,** such as Darwin's, as *'proper science'?* [Have we lost our minds?]

'Is this why we are now told that it is ***'unscientific'*** *to even* ***wonder*** if intelligent causation could play any part in the creation of reality?

'Is this why it is now *'unscientific'* for scientists to even ***wonder*** whether Life ***really*** could arise as a result of the wholly MINDLESS processes into which Darwin and his **materialist - atheist fans** ask us to put so much *faith?*

'That, according to wikipedia, to even *think* of doing so is a 'religious' argument? Clearly, it is an **atheist** argument to opt for Darwin's theory of 'unintelligent' design—that blind chance can create all. Which means it should ***not*** be taught in schools and colleges all over the world, that's for sure.

'Wikipedia makes yet more false assertions. It states:

> "Educators, philosophers, and the scientific community have demonstrated that ID is a religious argument, [not true] a form of creationism which lacks empirical support [not true] and [it] offers no tenable hypotheses." [Not true.]

'Just because ID hypothesizes that there is *more* to Life than mere chance, does not make it a 'religious' argument. A *super*-naturalist argument? Yes. Religious? No. Because, we can inquire into the *supernatural,* regardless of whether or not we have a religious position.

'Darwin's theory of pure unintelligent design, or UD, which wikipedia favors, is, *itself,* a quasi-religious argument, in that it rests on the utterly unscientific *faith* that BLIND CHANCE can create everything. Unscientific because it is not merely unlikely but demonstrably impossible. Further, if Living things *are intelligently* put together, we will *never* be able to *fully explain* them, if we keep pretending that they are not!

'Like pretending a jumbo jet *'must'* have been assembled by a tornado?'

'Yes. Wikipedia takes the stance that we already *'know,'* for a fact, that the *super* natural either plays no part in the creation of reality, or, that it *does not exist at all,* when these are the very questions needing answers.

'It is also not true, Ali, that "educators, philosophers and the scientific community" have falsified ID. Although it is clearly true that if we are *convinced* materialists or radical-atheists, we **have** to be opposed to ID—unless, that is, we have a powerful change of *mind and heart.*

'It is also wrong to say that ID 'lacks empirical support,' even as, ironically, it is very clear, by now, that Darwin's ideas—of (a) the lucky-chance combining of CHEMICALS into *Life,* and (b) its evolution, from there, by similar, blind-chance, means totally lacks empirical support,

'Whoever writes these articles in wikipedia, [at this author's time of writing] either does not understand ID, or, for philosophical or ideological reasons, positively dislikes it. Yet… one could just as well say,

> 'Many philosophers, educators and scientists have shown us that Darwin's theory of **unintelligent design**, UD, is a MATERIALIST (atheist) argument. [It is often used this way]. It is a creation myth for MATERIALISM [true] **which lacks empirical support** [very true]. And it offers no testable hypotheses [false].'

'While it is true that Darwin's *'Emperor's New Clothes' "Just So"* story of ranDom microbes to rAndom men—due to <u>aimless</u> chance, as wikipedia puts it—serves as a ***creation myth*** for the modern materialist and atheistic ***faith system***, it *can* be tested, so it's not beyond scientific inquiry.

'But it has been tested, very thoroughly by now, and, it has been found *very wanting,* by now.'

'Not not just once, Ollie, but again and again…'

'Oh dear me, yes! From genetics to molecular biology, from the fossils to embryology, Darwin stands falsified, but, like the foolish emperor in the *Emperor's New Clothes* story, we continue to be asked to agree that he looks magnificent, to consider ourselves intellectually satisfied with his, by now, *very* outdated, steam era fare, and to be grateful that he (supposedly) showed us that our existences are, essentially, meaningless—no soul, no purpose, no point.'

'What so concerns me, Ali, about *wikipedia,* on this, is that it is not like you or I putting something on the internet, and expecting people to agree.

'No, *wikipedia* is supposed to be a trustworthy source. It is not supposed to be a propaganda outlet for the currently fashionable, but intellectually vacuous, **materialist-atheist** world-view promoted by thinkers like Richard Dawkins and other famous atheists and materialists.

'On a topic as important as the science of origins, a source, like wikipedia, should, at the very least, allow BOTH SIDES of the debate to be put across properly—rather than taking such a partisan stance.

'Especially as, until Darwin's ideas gained such a weird hold on the western mind, the hypothesis of *spiritual* causation, or intelligent design, was always *mainstream* in western thought.'

We continued on our way.

21 Evolution as Religion

'In 1981, Colin Patterson, then senior paleontologist at the British Museum of Natural History, provoked an academic storm when he told an audience of fellow academics that Darwin's theory could no longer be considered knowledge but it had, like a religion, become a matter of *faith*. He said that one of the reasons he started taking a "non-evolutionary" view, was that he had "had a sudden realization."

> "One morning I woke up," he said, "and it struck me that **I had been working on this [evolution] stuff** for twenty years, and there was **not one thing I knew about it.** That was quite a shock, to learn that one can be **so misled** for so long. ...
>
> "So," he continued, "for the last few weeks, I've tried putting a simple question to various people and groups of people. The question is this: **Can you tell me anything you know about evolution**, ANY ONE THING THAT YOU THINK IS TRUE?
>
> "I tried that question on the geology staff in the Field Museum of Natural History, and the only answer I got was silence. I tried it on the members of the Evolutionary Morphology Seminar in the University of Chicago ... and all I got there was silence for a long time, and then eventually one person said: 'Yes, I do know one thing. IT OUGHT NOT TO BE TAUGHT IN HIGH SCHOOL.'" [230]

'Patterson continued,

> 'I woke up and I realized that all my life **I had been duped** into taking **evolutionism** as REVEALED TRUTH **in some way.** ...Now I think many people in this room would acknowledge that, during the last few years, if you had thought about it at all, you've experienced a shift from EVOLUTION AS KNOWLEDGE TO EVOLUTION AS FAITH. I know that's true of me, and I think it's

true of a good many of you in here. So that's my first theme. That **evolution** and **creationism** seem to be showing remarkable parallels. They are increasingly hard to tell apart.' ²³¹

'So you have Darwin's revelations, then—with no proof ever being found—you just have to *believe?'* Murmured Alisha.

'Yes. Darwin said that *On the Origin of Species* was just an introduction to his ideas, which he would later back up in a larger book, which, he never did. This is why, *without* the needed proofs, his ideas became a matter of *faith.*

'Remember, modern intelligent design theory is ***not*** the belief that any particular religious creation story is literally true. No. Just that ID does not disagree with the idea that a religious creation story may be *spiritually* true.

'While some do *try* to treat the old religious creation stories as if they are *literally* true, that is totally unnecessary and uncalled for.

'We do not need to know how something was *literally* created—be it a Cell, Fish, Human, or TV, Car, PC—to be able to work out that it was created by some sort of **purposeful intelligence**—not by mere luck. Remember, too, that to correctly ***describe*** the fossil evidence, which has been done very well by now, *is **not** the same as **explaining it.*** Our discovering that,

> 'This set of animals, with no obvious ancestors, ***suddenly*** appeared, ***did not*** morph into new types of creatures, as Darwin predicted, but ***stably*** remained the same, before, in time, ***dying out,*** then, later species mysteriously appeared in the same geologically abrupt way, remained the same, before also disappearing into extinction, only to be followed by further *Life Waves* of similar or different species . . .'

does not tell us what ***caused*** those sudden, ***non-evolutionary*** appearances of species. Our modern knowledge of geology belies the biblical-literalist idea known as young-earth creationism. But it gives no comfort to the Darwinian idea of a totally MINDLESS, ***yet highly progressive,*** evolution, culminating in the ***amazing, yet,*** BIOLOGICALLY TOTALLY UNNECESSARY, abilities of humanity: speech, poetry, art, music, science, mathematics.'

'Why "biologically unnecessary?"'

'Because evolution—Bacteria to Bears to Whales—is meant to be a totally MINDLESS process, isn't it? CHEMICALS, becoming 'alive' for no reason, then, starting to eat other chemicals! [Really?] Then, eventually, some of the **CHEMICALS** began to *think, to write poetry* and discover the laws of the universe. Why? For no reason at all! Ah, dear Ali, let us *think* about this:

'If the Darwinian view of evolution was true, why would "life" ever need to "progress" beyond some minimalistic, ***planet-covering, bacterial or algal stage?*** Why would it not simply continue, mindlessly, *"selfish-gene"* reproducing itself until the end of time—with no *"natural"* need whatsoever to *"progress"* beyond such a stage.

'So it should not be taught in schools all over the world as *fact?'*

'No, of course not. To tell young people that long ago, many weird and wonderful creatures arrived in the Cambrian Explosion, followed by further mysterious Life Waves, and mass extinctions, including the fascinating story of the intriguing dinosaurs, is reasonable, because it's what the fossils show.

'To admit that we don't currently know, scientifically speaking, what caused the diverse species to arise in the first place is also true. To concede that Darwin's theory does ***not*** accord with, ***nor*** does it explain the fossil evidence, ***nor*** the similarities in vertebrate limbs, ***nor*** embryology, ***nor*** homologies, ***nor*** genetics, ***nor*** the **DNA CODED** complexities of Living things, ***nor*** sex, ***nor*** the intriguing convergences and divergences between marsupial and placental mammals ***nor*** the convergence between the mammalian camera eye and the clever octopus's camera eye, ***nor*** the spider's amazing web, ***nor*** the bombardier beetle's extraordinary weaponry, ***nor*** the astonishing caterpillar-to-butterfly metamorphosis, ***nor*** vision, ***nor*** hearing, ***nor*** memory, ***nor*** the incredible biochemistry of blood, ***nor*** the heart, ***nor*** our amazing brains, ***nor*** lungs, ***nor*** AWARENESS, ***nor*** feeling, ***nor*** *thinking,* ***nor*** speech, ***nor*** *language,* is also reasonable.

'To admit that, currently, there are ***no*** scientific theories, that truly work, to explain ***any*** of these remarkable things would also be truthful.

'To explain to young people that some scientists are *spiritualists* and, others, are materialists, for whom, ***all*** living things are *unintelligently* designed, right up to their own and everyone else's *intelligent minds*—however self-contradictory that may seem, to you or me, would also be be fair. The young people can also learn that, while science is, at its best, a sincere search for truth, it can be affected, as the well known biologist, Richard Lewontin stated, by people's prior commitments to what they already believe to be true, or what they hope to be true.

'Those scientists, like Darwin, who believe that *only* the MATERIAL UNIVERSE is real, will put forward purely materialistic theories. While counter intuitive to most of humanity, they look at reality, at *living* nature, including, *(are we to suppose?),* even their own children and their own existences, as if it is *all mindless,* with **no meaning or purpose,** and this hypothesis, despite its many obvious flaws, can continue to be explored.

'Other scientists, currently, a minority, incline to the classical, *mind-before*-MATTER thesis. They adhere to the ancient intuition that the universe is *multi-dimensional,* that living things have *intelligent* causes, and that it may be possible—using modern, scientific means—to demonstrate this.

'Not that we need 'scientific' means to demonstrate it, do we, Ollie?!'

'I agree. Commonsense should be enough. Just looking at a butterfly, or bear, or whale, a wolf, eagle, or snail, an elephant or an infant, a sculptor, or a plumber, a rocket scientist or a trapeze artist should all be enough.

'While, clearly, some things in Nature are somewhat random, and 'unintelligently designed,' mountain ranges or river valleys for example, we don't need to be rocket scientists or biologists to work out that certain of its features, especially *living* Nature, and *bright* human minds, able to pursue *the arts and the sciences*, could never fall together by 'unintelligent chance,' no matter how much time we allow, or, no matter how philosophically annoying we may find it—be it because we hate religion or for whatever other reason.

'At the same time, it should be made clear to young people that contemporary hypotheses of *super* natural causation do not give credence to any particular set of spiritual or religious claims.

'As evidence supporting the modern, ID view, young people can learn of the cell membrane dilemma highlighted by Jacques Monod.[232] He wondered: 'How could the cell's very clever, selectively permeable membrane—which protects the DNA inside the cell from being damaged by what is outside—arise, absent the very DNA which CODES for it in the first place? They could learn that this, 'chicken or egg?' issue, is an example of **irreducible complexity**—which points to intelligent causality, not to chance.

'For those who want *scientific* proof that holistic reality is *intelligent, not 'dumb,'* we can mention cogent modern thinkers, like Behe, Meyer, and Dembski, who use powerful *scientific* techniques for working out whether certain patterns in nature are raNdOm or meaningful.

'For example, the Search for Extra-Terrestrial Intelligence (SETI) is a branch of modern intelligent design detection science, IDDS, as are archeology, (is that mound a natural feature of the landscape or is it artificial?), forensics, (was the deceased pushed or did s/he fall?), cryptography (is that gibberish or is it code?), and that SETI is based on the analysis of radio signals from space for signs of **meaningful non-randomness**.

'Intriguingly, researchers have realized that if portions of Life's **DNA CODES** were detected, by us, in the form of radio waves coming from space, the conclusion of **intelligence and artificiality** would be unavoidable.[233]'

'Is that because of the **specific** [non-random] **complexity** of Life's **DNA CODED** instructions: **TTAAGGCC + ATAGGAGC + CAACGGTT + CCTTGTAC + TATAGGCC + AGGAATGC**…?' Wondered Alisha.

'Yes, because it is not random or Shannon information. Researchers, like shCherbak and Makukov—having "noted **various indicators of artificiality** in the CODE, such that nucleon sums are multiples of 037, that the stop codons act as zero in a decimal system, and all the three-digit decimals (111, 222, 333, 444, 555, 666, 777, 888, and 999) appear at least once in the code, which also looks like an intentional feature,"[234]—write,

> 'NO NATURAL PROCESS can drive mass distribution to produce the balance … amino acids and SYNTACTIC SIGNS that make up this balance are *entirely* ABSTRACT since they are produced by translation of a string read across codons. … not only the signal itself reveals intelligent-like features [but]… **taken together all these aspects point at [the] A R T I F I C I A L nature of the patterns.**'[235]

'In these ways, young people would learn that true SCIENCE, unlike materialism, is not a rigid *belief* system. Rather, it's about keeping an open mind, and finding out what is *true*. For example, is Darwin's materialist

205

hypothesis of Life more likely true or the *spiritualist* hypothesis of more classical thinkers like Plato and Galileo, Newton and Planck?

'The fact that we don't currently know, scientifically, exactly how or why the stunningly complex functioning in living things arises, *second!, by second!, 24/7, our amazing hearts beating, right now!, and now!, and now!, our astonishing blood circulating, right now!, and now, and now,* nor exactly how, or why, the diverse plants, animals, and people arise, does not tell us that there are no intelligent processes involved.'

'So why not just give young people the facts, Ollie? "Life on Earth appeared billions of years ago, the Cambrian Explosion took place, trilobites appeared, spiders, snails, and fish, the dinosaurs came and went, birds and bats emerged," and explain that there is more than one way to interpret the data…'

'Yes, why not? For example, when learning that scientists have now realized that cells are *vastly more intricate* than Darwin and his **steam age** contemporaries imagined, and that, as this expert in cellular biology writes,

> 'We have **always underestimated cells**. … The entire cell can be viewed as a factory that contains an elaborate network of interlocking assembly lines, each of which is composed of a set of **large protein machines** . . . Why do we call the large protein assemblies that underlie cell function PROTEIN MACHINES? Precisely because, **like machines** invented by humans … these protein assemblies contain **highly coordinated moving parts**.' [236]

they can learn that: while many in science continue to claim that such complex processes could arise by unintelligent chance, not all agree.

'They can discover that there are increasing numbers of thinkers—like Milton, Denton, Johnson, Behe, Battson, Latham, Dossey, Sheldrake, Radin, Tour, Meyer, Axe, Gauger, Wells, Witt, Dembski, Flannery, Shedinger, Nagel, even some famous former atheists, like Anthony Flew—who doubt that these modern discoveries provide good evidence for 'purposeless' and 'unintelligent' processes in nature.

'Modern Galileos who question the *"Materialist Church's"* faith in St. Darwin's revelations that there are just *mindless laws + mindless chemicals + a "warm little pond" + countless (impossible) coincidences* and all Living Beings are mere accidents of these basic, but pointless, givens.

'So you think young people could cope with knowing that there are these two most basic of all the possible philosophies—materialism *vs spiritualism*—and scientists debate which is correct.'

'Yes, why not? We can explain that there are these two *very* opposed world views, and we can do our best not to mislead them into thinking (a) that physical scientists have the only valid answers, a delusion which is called **scientism**, (b) that it really *was* established, by Darwin, that Life is just an accident, which evolved by accident, nor (c) that 'science' has proved **materialism** (atheism) to be true and *spiritualism* to be false.

'What Colin Patterson highlighted with his question: '**Can you tell me anything you know about evolution that you think is true?**,' is that Darwin's ideas have gone from being once intriguing hypotheses—open to testing and to falsification, by the real world data—to a *faith* position.

'Darwin's ideas are now—despite their repeated falsification—treated like 'answers' to Life's mysteries which cannot be questioned.

'One odd thing, though, Ali. If you ever listen to a discussion about how science is *supposed* to be, you'll be told how scientists always keep their minds open to new ideas and to fresh evidence. That, *truly,* they are ever ready to revisit their earlier hypotheses and theories, and to humbly follow the evidence wherever it may lead. In many areas of science, this may be true, but, when it comes to Darwin, it seems not to be.

'We have seen—repeatedly—how the data doesn't match Darwin's predictions. Yet, the data is ignored, not once, but repeatedly.

'We have seen, how methodological materialism and methodological naturalism, MM/MN, while at times useful, can become science-blockers when they are used to close down perfectly reasonable lines of inquiry.

'We have seen how courageous modern Galileos—like Behe, Meyer and Dembski—are ridiculed or ignored. Yet, although Darwin's theory has failed to explain natural history, let alone *"The Origin of Species,"* which was it's whole point, relatively few seem interested. Isn't that odd?'

'So what's the answer...' Mused Alisha.

'Well, one answer is that our *own* purposeful desires, our own *amazing capacities* to do—*logical*—*intelligent*—*science,* ought to tell us that mindless chemical coincidences, in no matter how many 'warm little ponds,' cannot possibly explain the arising, in us, of the very faculties which we call *Soul and Mind, Intelligence and Curiosity, Will and Purpose, Experience and Memory, Imagination and Creativity,* all *non-material, spiritual* faculties which inspire us to pursue logical, intelligent, *Science!*

'It is only if we adopt **materialism** as our basic creed, that we will be forced to reject all notions of intelligence and purpose in Nature.'

'Which takes us away, Ollie, from an open minded inquiry and forces us into the *narrow, science-limiting, belief system* we call *materialism*. A belief system which makes us, and our science, meaningless!'

'Sadly. For, if, we can call a modern Jaguar Car or Jaguar Jet a cleverly designed object of beauty and purpose, but we are invited to pretend that its living, feline inspirations, vastly more clever, more beautiful, and more complex than anything we can make, using *all* our intelligence, are, merely, the **unintended byproducts** of blind **chemical coincidence**, which only *'seem' Soulful,* and Purposeful, and Alive, then our language, and our science, are no longer instruments for truth, but misleading tools for reality-distorting lies, and, whether we realize it or not, they have become *mephistophelian*.[237]

'But,' mused Ali, 'given that there has *never* been any really good evidence for Darwin's ideas, and that this was pointed out decades ago, even in his own day, why do people still cling to his failed ideas . . ?'

'Well, it's partly because, as Chris Carter points out,

> 'In our modern world, **science** and **scientists** hold a great deal of **prestige**, and few people want to be thought of as unscientific. ... [So] If to be scientific is good and unscientific bad, and if the term "scientific" is thought to be synonymous with the term "**materialistic**," then any talk of disembodied *minds* or *spirits* [or of intelligent causation in nature] is anti-materialist, unscientific, and therefore bad.[238] The long-standing **confusion** of materialism with science is what largely accounts for the persistent social taboo responsible for the ignorance and **dismissal** of the substantial amount of evidence that proves materialism false.'[239]

'"But," Carter adds, "the difference between science and ideology is not that they are based on different dogmas; rather, it is that *scientific beliefs* are *not* held as dogmas, but are open to testing and hence possible rejection." [240]

'Unfortunately, Alisha, MATERIALISM, for too many of us today, is not "open to testing and possible rejection." Its central dogmas:

>'Only the PHYSICAL is real. Blind chance is *the* great 'creative' principle. *Life* is just an accident, which evolved by accident. "You, your joys.., your memories and free will, are no more than the behavior of a vast assembly of nerve cells… You're nothing but a pack of neurons."' [241]

'Ah, this dreary ideology which asks us to believe so much nonsense just to avoid the far more logical conclusion that we are born of an **intelligent** *mystery,* which ***just Is,*** rather than a MINDLESS *mystery,* which ***just is****!*

'Unfortunately, for too long have we been presented with a *false* choice. Either we are to believe that the world was instantly created by a childishly simplistic, Father Christmas type god (false), or, it's far more ancient, (true), but it cannot be a product of intelligent causation at all (false). It all is pure accident, all the way up and all the way down. (false).

'Patterson was right. Nothing about Darwinian macro-evolution is true. "Darwin said so, a long time ago, so it must be right," is ***not evidence.*** Wikipedia saying: "scientists say that evolution is true," is ***not evidence.***

'Imaginative drawings of *imaginary* transitional species, never to be seen in the *non-imaginary,* real-world, fossil records, are *not evidence.*

'Darwin, for his time, had an idea. It was worth considering…'

'But it has not been borne out by the data…'

'Correct. Sadly, too many today, do not notice the glaring incongruence when Darwin's supporters say, "Finch beaks can change," 'Yes, we know that,' "Moths can vary," 'Yes, that's also true,' "and there are many different varieties of pigeon and dog," 'No problem, we all know these things.' But they don't stop there, oh dear me, no. They continue:

"So, this proves, as Darwin suggested to his friend Hooker, that DEAD CHEMICALS in a 'warm little pond,' struck by lightning, could turn into stunningly complex, DNA Coded, *Living,* self-reproducing, *Cells!"*

'***What?!*** But that's absurd. It's *chemically impossible!'* Mathematically, too, it's an absurdity.' Call out some, who are still paying attention. But, ignoring them, their materialist friends continue,

"Then, some of these *single* cells morphed into multi-celled worms, spiders and fish, and *sexual reproduction* started, by pure chance." 'Really?,' the still skeptical try asking. *'Has anyone shown* that this actually happened, let alone *how* it did?' Taking no notice, Darwin's fans press on,

"Then, water breathing fish, after undergoing yet more random genetic typo changes, became *air breathing* amphibians," 'How on earth..?!' A few, who have not given up, try asking. Yet, no answer comes back, except:

"Then dinosaurs, then birds, with *altogether different* lungs again, then mammals, and, eventually, brilliant minds like Plato, Newton and Einstein." A scheme which is not only unsupported, but *demonstrably impossible!*

'Isn't it odd that all this is not seen more critically?' Wondered Ali.

'Yes!, it is very odd. Changes over time? *Yes,* no problem. The mysteriously abrupt emergences of new species? *Yes.* Similarities in design? *Yes.* The process Darwin imagined? *Sorry, no.* An instant creation, 6000 years ago, based on Ussher's attempt to treat the western Bible as an early, but accurate, science text? *Sorry, also no.*

'So, all we really know,' started Alisha, 'is that Life on Earth is very old, and it has gone through many changes over vast time. Beyond this, it seems, we still know very little, scientifically speaking, about Life's true origins, what drove the mysteriously *sudden* appearances of new species, the long periods of *stability,* the eventual *disappearances,* followed, in time, by further, mysterious, Life Waves?'

'Yes. Currently, our sciences have no useful ideas as to what causes any *Living Being* to be, or *Mind, or Emotions* to arise out of the apparently MINDLESS DUST of this world, *this!* astonishing second, and *every! astonishing second, 24/7.* No one in mainstream, western, science today has any **useful idea** how *Cells!* or Breathing, Feeding! or Feeling, Sex! and sexual Reproduction, Female! and Male, blood!, and warmly pulsing Hearts!, all started or why our physical bodies slow down and, eventually, cease working, while, we, as immortal *Souls,* leave this dimension, at least for a while . . .

'The MINDLESS ATOMIC ELEMENTS of materialist belief are not *alive* in the way we are alive, nor do they need to *breathe!, or feed, or excrete, or sleep! or die* as we do. Yet our sciences are, currently, remarkably incurious about the true origins of all these fascinating qualities and functions.

'They have nothing useful to say about them. As they deny the inherent *intelligence, in the astonishing!, amazing, living!,* nature all around us, as they deny the *Life Forces!, the Chi, the Prana, the Soul!, the Spirit!,* in all things, including, even, in themselves, how can they?'

'You know something, I have noticed, Alisha, when it comes to thinking about Life's origins, today? It is that many people are, either unfamiliar with the classical notion of *'mystery,'* or, they actively dislike it.'

'Why? I *love* the word mystery. So many things are *so* mysterious . . .'

'Yes, but it makes them feel uncomfortable. They want all things to be worked out in totally LINEAR and MECHANICAL ways. Some call this approach, **reductionism**. Others call it, **scientism**, the worship of physical science, almost as if it is a religion, in the hope that, eventually, we will bring *all* of reality under our PHYSICAL, SCIENTIFIC & TECHNICAL control.'

'It's not a bad thing to want to understand things, is it?'

'No, of course not. Let us do our best to find out as much as we can about what is really true, about what ***really Is,*** including Life's origins.

'Yet, let us also never forget that, no matter *how* far we go in our researches, all we will manage to do is to penetrate just a little further into the *deepest* mystery of all, which is: '*Why* is there any-thing at all, rather than no-thing at all? *Why* is there a *final Source of All* which **Just Is**?

'Be it 'dumb,' as our materialist friends argue, or intelligent, as the *spiritually* minded believe? No one can say. The ***final Source of All***—be it mindless or be it ***Intelligent—just Is.*** This **Final Source,** be it, mindless and dead, or, ***intelligent*** *and astonishingly* **Alive!,** is the *deepest* mystery of all—and it is not a mystery which we will ever solve. Classical logic, useful as it is, **can never** tell us **why** there can be any **Being** or **Thing** rather than none.'

'***Why*** should there be ***any Beingness*** at all? Even one atom?'

'None can say. Yet . . . *there is* '**Being,**' *and* '**Isness,**' and, using logic, *and* intuition, we can attempt to make sense, both of its ***multitudinous parts,*** atoms, cells, plants, animals, people, planets, stars, and of its ***unitive totality***—be it a grand theory of everything, as *scientists* seek, or *a* ***direct mystical experience*** of the **Oneness** of everything, as the *mystics and yogis, the shamans and sufis* report to us. Which *"impossible"* of the seemingly *impossible* possibility of there being anything-at-all *vs* nothing-at-all makes more sense?

The 'mindless impossible' of MATERIALIST thinking?
Or, the '**intelligent** impossible' of classical *spiritualist* thinking?

'This is why, if we could just allow ourselves to admit that, currently, we do not have even the beginning of a whisper of a real idea, scientifically speaking, what *Life and Mind, Feeling and Needing, Love and Joy!, Curiosity and Compassion* truly are, nor what causes any *Being! or thing* to be so *magically Alive!,* and so *mysteriously Sentient* in this, *our enchanted, Soulful* world, we'd be on much firmer, if humbler, ground.

'The current *'science'* view, expressed by thinkers like Crick and Lewontin, Dawkins and Valtaoja,[242] 'solves' existence's **greatest mystery**, the existence of any.thing at all *vs.* no.thing at all, by proclaiming that:

> 'Sorry folks, there's nothing *truly* mysterious, *or* miraculous, *or* Intelligent, *or* Spiritual to see in regard to Life, to the very fact of **Existence-Being** *vs.* non-being. It is all just 'dumb-lucky' chemistry, with no more *'Soul'* or 'Intelligence' to it than there is to a randomly evolving river valley. There is no more 'meaning' to it all than we care to make up. This, folks, is the 'rational' view, the 'scientific' view. Now, go out and enjoy your meaningless existences!'

'This is the impossible, unrealistic, inquiry-limiting, science-crimping, reality-shrinking story so many now believe in—in no insignificant part thanks to Darwin and his friends.

'Yet, as to *why,* let alone *how,* the Soulless, the mindless, the insentient little ATOMS of the materialist *faith* would ever become Beetle, or Eagle, or Lion, or People shaped, let alone begin to *Think,* and to *Create,* and to *Soulfully Care* about anything at all, let alone be able to *enjoy, to enthuse, to be curious,* to be able to seek *meanings* and to pursue *rational,! meaningful, scientific!, philosophical,* and *spiritual* inquiries, our current sciences have little useful to say.'

We continued on the beautiful riverside way.

Part II

Beyond Materialism

Spiritual Science

Modern Research into, and
Evidence for, the
Spiritual

22 No One to Know, Nothing to Know

'That *anything* exists at all, is the deepest mystery of all. But is it an intelligent mystery or MINDLESS? Darwin inclined to MINDLESS. But his ideas, which I adopted as a teen, that there is *no inner or Higher Nature, no Soul,* that all is RaNdom and purposeless, no longer make sense to me...'

'What, after all, is *truly* causing our minds to *think,* our hearts to *beat,* and our lungs to *breathe,* right *now,* this *amazing second!, and every amazing second!?* Does anyone *really* believe it is all due to an absurdly implausible series of impossibly unlikely chemical coincidences? Is this way of thinking, *honestly* intellectually satisfying—let alone scientific?

'That our abilities to ***imagine*** and ***to feel,*** to create art and to discover mathematics are all just MINDLESS coincidences of MINDLESS ATOMS, tossed around in a MINDLESS cosmic sea, until they randomly bumped-and-clumped together and, by MINDLESS chance, became stunningly complex microbial life, and, eventually, *living, thinking,* people—like you and me?

'It does seem strange, Ollie, to use our, *own, bright minds* to say that the *amazing, Living Nature* which gives birth to us is, itself, as *dumb-as-dust...*'

'*Yes!* There are ***zero*** evidential grounds for claiming that mere luck could drive an unfoldment of *Life,* in a progressive way of ever rising 'improvement,' as Darwin put it, leading to the *human mind* and all its amazing, yet ***biologically totally unnecessary,*** capabilities, other than:

> "Now we realize that the world wasn't, *literally,* made by a 1 heavily bearded, Father Christmas type god, out of spit and clay, in just six very busy days, what other option is there?"

'What a failure of the modern understanding. Contemporary ID thinkers, like Denton, Johnson, Lennox, Behe, Meyer, Dembski, Witt, Wells, Latham, who explore intelligent design, *vs.* Darwin's theory of unintelligent design, are not religious literalists who try to treat the scriptures as *'science.'*

'No, they really are not! All they're saying is that just because it is easier to explore OUTER, nature, and more challenging to inquire into *Life's* deeper causes, we should not give up and assume that *Existence-Life-Force-Source-Being, has* no deeper sources. If any materialists, well known or obscure, were prepared to listen to me, Ali, I'd love say to them:

"Look, you don't have to belong to this or that religion, or believe in some absurdly simplistic 'bearded god,' making and sustaining everything, second by second, but can you not conceive that all of **This,** *tiny quark to mighty galaxy,* may be the expression of an *amazing, all pervading intelligence* which ***just Is***—call it God, Buddha Nature, or the Tao, call it what you like, the name doesn't matter—rather than a random, cosmic MINDLESSNESS, which *just Is,* a soulless, pointless, RANDOMNESS which *just Is?*

"Where is the reason in your claim that our *bright* minds 'raNdo*m*ly' emerge from mindless COSMIC DUST, in pointless combination, descended *'by chance,'* [what chance?!], from an *'accidental'* microbial cell—itself more complex than a super-computer and spaceship[243] combined—rather than the classical knowing that we *meaningfully* emerge from a *multi-dimensional* causal matrix which, to put your minds at rest, looks nothing like *Father Christmas!* Wouldn't that be a pleasant release from your unreasonable Darwinian paradigm of utter unmeaning?"

"But, where would such an *Intelligent Energy, Law, Force or Source* come from? Who'd make it? Who'd design it?" They cry, echoing Dawkins: "Who 'designed' God?," he once asked.

*"No one! This stupendous Intelligence, this Great Spirit, this mysterious Force, this Tao, this call-it-what-you-like, **Just Is**...* Who or what, after all, generates all the mindless, random, happenings, you fill your knowledge G A Ps with, and which you believe to be so amazingly creative?"

"No one, they *just mindlessly **Are!**"*

"But **where** do all your mindless chances come from?"

"Well, they must just come from some kind of Mindless, Unintelligent and Accidental Universe Generating Mechanism, or MUA-UGM, which, very mysteriously, if pointlessly, *just Is.*"

"Ok… but *where* does your MUA-UGM come from? Who or what caused it to, so amazingly, if so pointlessly, *just Be?*"

"*Who knows?! It just IS!* Beyond the mysterious MUA-UGM, required to generate all the pointless, random events, which eventually lead, by pure stupid chance, to nature's astonishingly precise laws, to her smart materials [carbon, hydrogen, oxygen etc], to the fascinating *Living World,* to our *bright* inquiring *Minds,* and to our sometimes caring Hearts, there is no possible regress."

Alisha smiled. She said, 'So this hypothetical MUA-UGM functions as a logically necessary pseudo-god or *Final Source* and *Uncaused Cause*—which *Just Is*—for the modern materialist-atheist belief system?'

'Yes. Materialism argues that we emerge from the de-Souled and de-purposed matrix of its dry faith, imbued with amazing *Awarenesses* and clever *Minds,* full of *Purposes, Loves, and Desires,* but that, really, our *Creativity and our Ideas, our Dreams and our Plans,* are, in the final analysis, all just MINDLESS coincidences of MINDLESS DUST!

'Yet the MUA-UGM is intellectually hopeless—not satisfying, at all. It has no *will,* no *desire,* no *intelligence,* no *creativity.* If it could account for anything, it would only be for *half* of what exists, for MATTER and MECHANISM, the 'TAILS' sides of the coin of reality, *not for Mind or Soul.*

'But mechanisms, Ollie, always mean *functioning,* which always points to *purpose,* because mechanisms always function for a *reason,* which points to intelligence(s), *with* reasons, which points to *Subjectivity and Spirit.'*

'MINDLESS universe or **meaningful.** Which makes more **sense?** *We* are intelligent, *we* have minds. Is it not more reasonable, then, to assume that *the stupendous Existence,* from which we emerge, is also *Intelligent*—in its *very Nature,* from tiny atom to mighty **galaxy**? Could not **what-Is-ness,** be *intelligent,* in its very *ontology,* as the classical philosophers inferred?

'Ironic, too, that materialism consists in `immaterial ideas,` and that, as Max Planck explained, "**There is no matter as such**[!]. All matter originates and exists only by virtue of *a force ... mind ...* mind is the matrix of all matter."[244] A view echoed by the physicist Sir James Jeans, 'the Universe begins to look more like a GREAT THOUGHT than like a GREAT MACHINE. Mind no longer appears to be an accidental intruder into the realm of matter . . . we ought rather hail it as the creator and governor... of matter.[245]

'If we think about it, Ali, without *Sentiency or Mind,* there cannot *be any Existence* or *Being. How* can there be any *'thing,'* or *'being(s),'* or *'beingness,'* of any kind, without a *Knower or Knowers of knowing Being* to know *knowing Being?* It even takes *Mind, mine, yours, or anyone's,* to imagine a universe *without* Mind...

'I guess, Ollie, many still try to look at the world as though it can exist *independently* of Mind, as though it is all FORM, OBJECT and MECHANISM, as though it has *no Subjective, Soul, or Spirit* dimensions.'

'Yes. Yet, as India's yogi sages long ago explained, there cannot *be Existence, or Being,* of any kind, without *Living, Sentient, Subjective, 'I am' Being* to know Itself, and, to *in.*FORM *its-Self—to bring its-Self into* FORM.

'We cannot dispense with the `mental`: it takes *Witnessing Awareness—* which implies *Self or Subjectivity,* that is, **Knowing**, **Living**, **Beingness**—to **Know**, **Living**, **Knowing Beingness**. Without witnessing **Awareness** or pure **Knowing Being**, *to know* knowing **Being**, expressing as **what-Is-ness**, there cannot *be Being* or **what-Is-ness**, there can only be mindless **oblivion**.

'It is all pretty strange, though, isn't it?' Mused Alisha.

'Well, existence *is* strange. The very *is-ness* of things, and our very capacity to know anything at all, are the deepest mysteries. After all, hasn't the (seemingly) impossible happened? That there is anything-at-all, rather than nothing at all, would seem to be impossible, wouldn't it? *How* can there *Be* any *thing?* Let alone *a Cause* of any *thing?* Never mind *Universes!*

'Isn't it easier to imagine there being *no-thing* and *no-being(s)* to know it, rather than any *being-thing?* Yet, the (seemingly) impossible *has* happened. **Stars** and planets, **cells** and plants **all exist** . . . animals exist . . . **we exist.**

'While we may **never** know **how** all this is possible, we can ask which of these two [im]possibilities makes more sense? (a) *An Amazing, Blazing, Living, Loving, Light, Power, Will, Creative,* **Intelligence**, *which* unfathomably *just Is?* Or (b) a sterile, cosmic mindlessness which unfathomably *just is?* Either way, *Being-Existence-Isness* is deeply, infinitely, mysterious; and, clearly, creation is *not* some one off thing. It is happening *this very second* and *every* second . . . As Peter Wilberg says,

> "According to quantum physicists such as Hawking, while it may seem that your television or computer came from a factory in China, **in reality its every atom and particle is constantly manifesting from an invisible but all pervasive quantum 'vacuum.'** ... the mobile phone you pick up or the television... you look at today is not the one built in a factory–or even the one you picked up... a nanosecond ago! **For matter** – all matter – is constantly emerging from and disappearing back into a **quantum vacuum**.
>
> "What exactly this quantum 'field' or 'vacuum' is no physicist can say. Yet ... **It was said by those ancient sages, [the yogi-adepts of India]** who declared that every aspect of our experienced world – and everything in it – is constantly and continuously manifesting from a vast and infinite field of CONSCIOUSNESS *[God,Spirit,Tao]*." [246]

'Wilberg's points *are* in line with modern physics. He continues,

> '...the truth is that it is no mysterious 'quantum void' but rather *consciousness itself* which is the sole reality behind and within all things ... C O N S C I O U S N E S S as such, understood as an infinite, all pervasive field of pure *A W A R E N E S S,* one **latent with infinite potentialities** of expression that are constantly manifesting as every existing thing we experience – from a rock to a laptop. For as a saying of the great 10th century Indian sage Abhinavagupta expressed it so well: "**the being of all things that are recognized in** *A W A R E N E S S* **in turn depends on** *A W A R E N E S S.*" [247]

'Mysterious? Yes. That there is anything-at-all, rather than nothing at all is *beyond* amazing. It is, seemingly, *impossible.* ***How,*** after all, can there be anything at all, rather than noting at all? ***How*** can there ***Be*** any *thing,* **Atom,** *Cell,* **Planet,** *Star,* ***Galaxy?*** Let alone *a Cause* of any *thing?* Never mind

Universes! Isn't it easier to imagine there being *no-thing* and *no-being(s)*, rather than any *Being or Thing?* Yet, the *(seemingly)* impossible *has* happened. **Stars** and planets, **cells** and plants, **all exist**. . . animals exist . . . **we exist***!* So, which of the two [im]possibilities makes more sense? (a) A cosmic *intelligence, which **just Is***? Or, (b) a cosmic mindlessness which ***just is*?**

'Either way, *Being-Existence-Isness* is deeply, infinitely, mysterious…'

'Materialism say: "The cosmos is mindless—no *Soul, no Spirit,* no point. It *is* generated, *this* second, and *every* second, 24/7, by a mysterious, but, totally **Mindless, Unintelligent and Accidental Universe Generating Mechanism** or MUA-UGM which *'just Is.'* By unintelligent chance, it gives rise to *bright, intelligent* us, who, then, oh so 'cleverly,' work out just how empty and pointless it all is, while marveling all the while, in unnoticed contradiction, at our own cleverness."

'That this doesn't make the best sense of the data, especially of *Living things,* and *Living Beings,* let alone of C O N S C I O U S N E S S, and Mind, of Feeling and Thinking, of Curiosity and Creativity, is, strangely, ignored.

'Yet, to try to make itself seem *'true,'* this sterile, impossible philosophy has to deny all the [*obvious*] evidence for intelligence and design in Nature, while 'skeptically' dismissing every, single, *mystical, psychical or spiritual* experience anyone has ever had. Not noticing that Darwin's theory is *not* borne out by the data, that Life could *never* start by chance, nor *'evolve'* by chance, that people are *not* (always) deluded about their spiritual experiences, that tell us, along with the findings of modern physics, and the classical testimony of the prophets and the yogis, the shamans and the sages, that **'SOLID'** matter is not primary, *Mind and* C O N S C I O U S N E S S are.

'Not noticing that it has to disregard, what we now know for clear scientific fact, that, in *deep* reality, our world, which *seems* so **SOLID** to us, is not, truly, S…O…L…I…D at all, for, as Max Planck pointed out, *there is* **NO MATTER** as such, there is just S . . . P . . . A . . . C . . . E . . . 99.9…9…9…9…9…9…9…9…9…9…9…9% of it. It has to ignore that, at its depths, our apparently so **SOLID** physical r…e…a…l…i…t…y is ultimately made of *Shimmering, Living Fields* of *intelligently* in.form.ed e~n~e~r~g~y S…P…A…C…E…'

We continued on our way by the sparkling river.

23 Spiritual Science

'In *Mind and Cosmos,* the distinguished philosopher, Thomas Nagel wrote,

> 'If materialism cannot accommodate consciousness and other *mind-related* aspects of reality, then we must abandon a purely materialist understanding of nature in general, extending to biology, evolutionary theory, and cosmology.' [248]

'Our sciences, Alisha, for so long focused on matter and mechanism, ignoring the *subjective* and the *spiritual,* have *described* many things, but *not* fully ***explained*** them. To describe what a Life Form is *made* of and how it *works,* clever as it is, doesn't tell us how it arose in the first place.

'But the neglect of Aristotle's classical insight that all things have *not* only outer MATERIAL causes, but *formal and final* causes, shaping ideas and purposes, explains why we are in such a mental muddle today.

'Materialism limits our inquiries and cuts us off from half of reality. The truth of all that is *Subjective, Soulful and Intelligent* in Nature is denied.

'This narrow, soul-denying philosophy, created, ironically, by human *Souls,* has led our sciences to become so closed to the true *Life,* and *Soul,* in all things, not just in people and animals but even in *Gaia,* our *amazing!, living! Earth, Anima Mundi!*

'This is why our current sciences cannot imagine that our beautiful, magical planet may, in her own way, be a fascinating expression of *mysterious, multihued Being, enSouled and enSpirited,* as the world's earlier peoples knew, some still know.

'No, today, she's just a random rock. All her rivers and oceans, her forests and her mountains, are supposedly DEAD and *Soulless* too, because *'science says,'* there is no *Soul,* or *intelligence,* or meaning to anything.

'There are just ACCIDENTAL CHEMICALS having chemical accidents and highly *intelligent* and *purposeful* scientific `minds (Souls!)` telling us just how 'unintelligent,' and 'purposeless,' and 'accidental,' it all is.

'So, Ollie, for this uninspired way of thinking, *Gaia* cannot be our immediate, *Living* source, a vast `Consciousness` in her own right, Mother Earth in a *Soulful* Cosmos of intelligence and meaning? *Pachamama,* as the Andean peoples call her? She's just a random rock, even though, as James Lovelock explained,[249] like all Living organisms, she has many homeostatic [self-regulating] features which make Life possible.'

'Yes, but, sadly, it is not just materialists who reject this ancient intuition, but some of the major religions too. Why? Because they think of this ancient intuition as nature worship, of which they disapprove.

'Regrettably, they seem to think that the *God or Supreme Creative Spirit* of their beliefs would give them a **dead** world, a **de-souled** world, rather than a *Living—enchanted—en-Souled* world—*Anima Mundi.*

'What happened to the *Living Soul* and *Living Spirit* in our deeper understandings? Do these great religious traditions not see that it was the denial of the *soulful* and the spiritual *within* nature, and the persecution of those they dismissed as animists and pagans, the world's first peoples, which laid the foundations for today's dismal spirit of materialism? This dreary ideology which—in the end—denies the *Soul* and the *Spirit* altogether.

'Some people think that if we acknowledge the *Soul* and *Spirit* in the mountains and the rivers, in the woods and the trees, in the great oak and the giant sequoia—there, long before we were born, and long after we are gone—we are being fanciful or animistic. But, if *Soul* and *Spirit* are real, we are merely acknowledging the *inner, Living Beingness* in all things.

'In his near death experience, after he was struck by lightning, with his consciousness floating, momentarily, outside of his body, described in *Secrets of the Light,* Dannion Brinkley reports,

> 'As all this transpired, the vibrant kaleidoscope of living colors that emanated from Tommy and Sandy [trying to resuscitate me] astounded me. In fact, as I glanced around the room **everything appeared to be literally alive and vibrating with color.**
>
> 'Even the wooden chest of drawers in the corner radiated a multihued energy. What an amazing observation for a redneck! I wish everyone could see what I saw on that night. It certainly

changed the way I relate to animate, and even inanimate, life to this very day. I no longer take for granted the unique and beautiful spiritual life force flowing through every creation in the physical world. To witness how intricately everything was connected, interwoven, and related, at the deepest levels of a highly organized matrix of networked energy, was indeed an overwhelming and inescapable new reality for me.' [250]

'As Dannion found, the *spiritual* and the *supernatural,* the *Life forces* and the *Soul,* are **all around us**, not just in other dimensions, next door.

'They are in the animals and the fields, in the children playing, and in the wind waving leaves. The forests and the mountains, the rivers and the seas are all alive in their own way.'

We walked on, by the river, golden cattle gently grazing.

'Physical science, Ali, is based on physical measurements and, when circumstances permit, repeatable experiments. On the other hand, what some refer to as *'spiritual science,'* of which the world's esoteric traditions, like yoga, shamanism, sufism, theosophy and anthroposophy, are parts, is not based solely on the data of obvious, outer reality but also on data obtained by *inner* vision and *extra-sensory* perception, faculties latent in all of us, faculties which we can all develop, in principle, if we wish . . .'

'Go on,' said Alisha, looking intrigued.

'Well, there have always been those who have been able to see, to some greater or lesser extent, into the *subtle* dimensions of reality—either by virtue of natural gifts, *spontaneous* awakenings, or by *inner* training, as some psychics, yogis and other mystics undergo, or, indeed, *anyone* with sufficient interest and the right aptitudes can undertake. Rudolf Steiner, the famous Austrian philosopher, clairvoyant and spiritual-scientist, wrote,

> 'There slumber in every human being faculties by means of which he can acquire for himself a knowledge of higher worlds. Mystics, Gnostics, Theosophists – all speak of a world of soul and spirit which for them is just as real as the world we see with our physical eyes and touch with our physical hands. At every moment the listener may say to himself: that, of which they speak, I too

can learn, if I develop within myself certain powers which today still slumber within me.' [251]

'Such people, high-level yogis, mystics and psychics, have seen into these deeper realities and reported their findings. Many otherwise everyday people have also had powerful *psychic* and mystical experiences, including precognition, telepathy and out of body experiences. They provide a lot of data for those researching into what we can refer to as *Spiritual Science.*

'For example, in the late nineteenth and early twentieth centuries, the well known *theosophists,* Annie Besant and Charles Leadbeater, explored the sub-atomic world using extrasensory perception or ESP.

'In his book, *The Secret Life of Nature,* Peter Tompkins describes how he set out to see if there "really was an acceptable correspondence" between Besant and Leadbeater's amazing yogic descriptions of atoms and 'the 'reality' of orthodox physicists.' He wrote,

> 'To find out I went in search of the first qualified theoretical physicist to reevaluate the theosophists' pioneering work in ***Occult Chemistry***,[252] Dr Stephen M. Phillips, **a professor of particle physics**. Phillips's book ***Extrasensory Perception of Quarks***, published in 1980, while dealing with the most advanced nuclear theories, including the nature of quarks, postulated particles even smaller than quarks as yet undiscovered by science.
>
> '**Analyzing twenty-two diagrams of the hundred or so chemical atoms described** in *Occult Chemistry* . . . at the turn of the century, Phillips found it hard to avoid the conclusion that "Besant and Leadbeater **did truly observe quarks using ESP some 70 years before physicists proposed their existence**". What is more, their diagrams indicated ultimate physical particles even smaller than quarks.
>
> 'By the time I discovered Phillips on the southern coast of England…he had checked another eighty-four of the theosophists' atoms: all were seen by him to be **100 percent consonant with the most recent findings of particle physicists**. Every one of the **3,546 subquarks** [!] counted by Leadbeater in the element of gold could

be correctly accounted for by Phillips. Were Phillips's conclusions to be substantiated by his peers, it would adduce evidence that the theosophists with their yogi powers had **effectively opened a window from the world of matter into the world of spirit**.

Prompt and committed approval of Phillips's conclusions had already come from the noted biochemist and fellow of the Royal Society, E. Lester Smith, discoverer of vitamin B-12. At home in both the mathematical language of physics and the arcane language of theosophy, **Smith spelled out his support in a small volume**, *Occult Chemistry Re-evaluated* [1982]. And Professor Brian Josephson of Cambridge University, a Nobel Prize winner in physics, was **sufficiently impressed by Phillips's radical thesis to invite him to lecture on the subject** at the famous Cavendish laboratory in 1985.

Yet few in the ranks of orthodoxy **had the courage to risk their positions** by supporting anything so wild as the notion that **psychics could see [just as well or] better into the basic constituents of matter** than could physicists armed with billion-dollar supercolliders.' 253

'That's intriguing, Ollie.'

'Yes, and it's very important. *Why?* Because science depends on good quality evidence, on sensory evidence. Materialism claims that we only have physical senses which arose—for no reason—by pointless chance, to pointlessly report, to a pointless 'CHEMICAL-ACCIDENT' 'SELF,' not to a *true Soul*. But, if there are other *subtle* senses, which we can use to gain reliable knowledge, not only of *spiritual* nature but, even of mundane, physical nature, then this is another way in which this unrealistic paradigm is refuted.

Leadbeater and Besant used such *other* senses. THEY FALSIFIED MATERIALISM. Their work should have been epoch making...'

'Okay, so if Leadbeater and Besant used trained *psychic* perception to accurately describe sub-atomic particles, some time before conventional scientists even *knew* of their existence, why isn't this more widely known?'

'Because, when their book, ***Occult Chemistry***, came out, a hundred years ago, no conventional scientists HAD SEEN INSIDE AN ATOM.

'So, they had no way to evaluate the *theosophists* descriptions?'

'Yes. It wasn't until much later, when particle physics expert, Dr. Stephen Phillips, came across a copy of *Occult Chemistry*, that their fascinating accounts of the particles could be verified. Peter Tompkins writes,

> 'Phillips counted the quarks, and saw that **the Theosophists had the correct number of quarks in every element** – and the last quark was only discovered in 1997 – the proof of it is self-evident.
>
> 'The excuses for disbelieving the claims of the psychics are irrelevant in the context of their **highly evidential** descriptions of subatomic particles published in **1908**, **2 years before** Rutherford's experiments confirmed the nuclear model of the atom, **5 years before Bohr** presented his theory of the hydrogen atom, **24 years before Chadwick** discovered the neutron and Heisenberg proposed that it is a constituent of the atomic nuclei, **56 years before Gell-Mann and Zweig** theorized about quarks. **Their observations are still being confirmed by discoveries of science many years later.**'

'Professor Phillips himself published a book, ***Extra–Sensory Perception of Quarks*** [254] confirming his findings. But, according to Tompkins, although his book was very interesting, it was,

> 'too scholarly – too much mathematics and physics in it for the general public. [And] For [many] scientists themselves, it **shatters their whole [materialist, anti-supernaturalist, anti-spiritualist] premise** and if they don't want to look at something, they won't. In fact they will do their best to put you down . . .'

'In an interview, Tompkins was asked whether the fact that modern physics has confirmed the theosophist clairvoyants descriptions of sub-atomic matter means that their *super* sensory descriptions of the *spiritual hierarchies,* the worlds of the angels, nature spirits, devas and fairies can also be taken as real? He replied:

> 'If a hundred years ago the Theosophists were absolutely correct in the number of quarks which they described, then one has perforce to look at their detailed descriptions of *nature spirits*.

'Obviously they are using effective clairvoyance. When you find that the **shamans all throughout South America, for instance, describe the same phenomena as the shamans in the Far East**, rationally you must take it into consideration as being possibly real.

'Why would people accept the physicists' description of an atom, which the general public cannot see, and not accept the description of the nature spirits that they cannot see? They accept the religious notions of spirit [and spirits], though they cannot see them, **why not accept the possibility of nature spirits and [the] angelic hierarchy?** . . .

'The whole Christian persecution of witchcraft, and the [denial of the] whole world of nature spirits, is what has alienated us from a healthy planet. It's only when we get back in touch [with] and accept their presence that the whole thing falls back into place and we take our place in the cosmos. **Otherwise we're totally alienated from a healthy planet** and from our role in the cosmos.' [255]

'You see, Ali, while it seems hard for many of us moderns to understand, the world's earlier peoples realized that there is not one mineral, flower, tree, animal or person, not a single forest, mountain, river or sea, no planet, star, or galaxy, of total holistic reality, which does not have *inner, subjective, or spiritual* dimensions. They realized that deep reality is *utterly alive*. It consists in many different levels of consciousness, it is *multi-dimensional*.

'In their bestselling book, *The Secret Life of Plants*,[256] Tompkins and Bird showed how the Plant world is **pervaded** by intelligence and subjectivity. They described the wonderful sensitivities of plants, their capacities to be used as lie detectors, their fascinating abilities to help each other, and to adapt to human wishes, their intriguing responses to different kinds of music, their ability to communicate with humanity.

'Other thinkers, like the formidable Blavatsky, founder of theosophy, and the titanic Rudolf Steiner,[257] founder of anthroposophy, biodynamic agriculture, Waldorf education, anthroposophical medicine and eurythmy, have also averred that all dimensions of reality—even the mineral—have *subtle, sentient* elements, as the world's first peoples always said.

'These people all take a *panentheistic* view of reality, which is to say that an *All-pervading, Intelligent, What-Is-ness, an It/Is/I/S/he or ISH,* is not just *'out there'* in the *beyond* but is, also, **right here**, closer to us than many of us ever realize. Not separate from the Eons-Long-Creation-in-Evolution (ELCIE) but the **OUTER FORM** and *Inner Life* of every thing.

'Every last quark, electron and photon of total holistic reality is a meaningful vector for, and vibration of, this amazing *'what-Is-ness-being.'* Atoms vibrate at 10,000,000,000,000 cycles per second—that is **very** *alive!'*

'Are you saying then that the whole universe is **alive?***'*

'Yes! Multi-dimensional reality consists, *ultimately, in Living-Spirit-Being.* India's yogi-adepts talk of it as *Living–Creative–Being–Consciousness.* This is why they say: *'Tat Tvam Asi,'* '***That*** *you Are.'* Which is to say, 'You, too, are *woven* from the very *stuff* of *self-existent,* 'I AM THAT I AM,' *Beingness—you* are made in its image.'

Alisha spoke quietly. She said, 'This intuition that *all is Alive, Living Presence, Essence, Intelligence, Soul, Consciousness, Spirit, Being,* even down to the plants and minerals, is why it's so important to treat *all* of creation with as much respect as possible, not just grub, grab, stake, take, make, and break, but rather to say, 'please and thank you,' as the world's earlier peoples did, some still do . . .'

'Yes, that, my dear friend, is so true. This is what Steve Taylor says about this, in his fascinating book, *Waking from Sleep.*

> '. . . to indigenous peoples there are **no such things as "inanimate objects"**. . . . they sensed that even rocks, the soil and rivers had a *being or consciousness* of their own. Whereas European or Asian peoples tend to see these things as nothing more than one-dimensional objects which they are entitled to use for their own devices, **indigenous peoples** felt that the natural phenomena around them were **alive** and so treated them with respect.. . .
>
> 'As anthropologist Robert Lawler notes, "Every distinguishable energy, form or substance has **both an objective** and a **subjective** [*Soul or Spiritual*] expression." Aborigines have

complained that the problem with European Australians is that **they can only perceive** the surface reality of the world and ***can't enter its interior life***. As one aboriginal elder stated,

"Unless white man learns to enter the dreaming of the countryside, the plants, and animals before he uses or eats them, **he will become sick and insane and destroy himself**.'" [258]

'Unlike us, Ali, those we now think of as the world's *first* peoples, along with the yogis and the mages, the shamans and the sages, realized that holistic Nature is both *a unity,* or *non-dual,* yet, like a coin, which is *one-thing,* it can be considered, as Descartes suggested, *dualistically.* As: *Subjects* (*Heads*) and OBJECTS (TAILS).

'*Subjective Mind vs.* OBJECTIVE[259] MATTER, *Subjective Life vs.* OBJECTIVE FORM, *Subjective, Inner, Souls* vs. OUTER MECHANISMS, *Spiritual* vs. PHYSICAL, *Intelligent instructions (Life's Codes) vs.* OBJECTIVE, CHEMICAL INK (DNA: A.C.T.G.)—which are the PHYSICAL CARRIERS of that *abstract, non-physical software / information*.

'They realized that we are not only *part* of the *'All That Is,'* but we are *made* of it, and that, as microcosms of the macrocosm, we are blessed with the magical possibility to *know* it. Originally, science began because people, noticing the orderly movements of the planets and the stars, and the *wisdoms* and *intelligences* embodied in the plant, animal, and human kingdoms, assumed that these had `intelligent,` not mindless sources.

'After Darwin, sadly, the parts of our science relating to *origins,* devolved into an activity in which *purposeful, intelligent* people devoted themselves to explaining just how *'unintelligent'* and *'purposeless'* Nature is, without noticing the glaring contradictions in their own positions.'

'But their argument, Ollie, is that the *subtler causes* of reality are not physically visible.'

'Yes, but we need to *look* around. We are *surrounded* by endless expressions of beauty, Soul and intelligence made *visible*. It should be utterly obvious to us that there is far more to *Life, and Mind,* than mere chance, and most of us have experiences which point to *deeper* realities.

'There are, basically, *just two* ways to determine whether intelligent causation plays any part in the creation of reality or whether all is mere chance. Firstly, logic and the laws of probability, which tell us that living things could not arise by chance, it's not merely unlikely—it's *impossible.*

'Secondly, our *intuitions,* including, at times, *spiritual, and extra-sensory experiences,* which tell us that Living things do not *look* unintelligent or accidental because they really are not. That when we look at a new born baby, or at the soft, fresh buds on the trees in spring, we are not looking at mere chemical accidents, but at *(true) miracles,* unfolding before our eyes.

'Lesser miracles within the *Greater Miracle* which it is to weave the *Sacred Tapestry of Existence* with *Self-Expressed* cosmic silk and wool, from no-thing to some-thing, from no-where to now-here. A view endorsed by the world's prophets, yogis, shamans and other mystics, together with the testimony of many perfectly ordinary people, who confirm these things.

'Of course, some people, like the world's first peoples, the *mystics* of all faiths, and modern *psychics* or *spiritualists,* have always known that the universe is not only material but also *spiritual—multi-dimensional.*

'Modern physics also tells us that the universe is not, ultimately, 'material' *at all.*[260] Rather, it is in.FORM.ed, brought into form, through high vibrating energies, which, ultimately, disappear into f...o...r...m.. L...E...S...S... n..e..s..s.., a shimmering *Cheshire Cat's Smile* of a Divine Dream, full of magic and meaning.

'Bring physical science and *spiritual* science together—spiritual science, being the exploration, where possible, of the *non-physical* dimensions of reality—and we get a more complete picture of the holistic (non-dual, yet, also, dual) 'coin' of nature's two halves: OUTER (TAILS), MATTER and MECHANISM, from atoms to galaxies, and everything in between, and *Subjective, inner, Spiritual* nature (*Heads*).

'All of it an expression of *Living–Creative–Spirit–Intelligence, (call it God, Allah, Buddha Nature, or the Great Mystery,* the name doesn't matter), hologramatically self-fragmented into trillions of lesser sparks, diamonds and flames of C O N S C I O U S N E S S, expressing as *all the subjectivities,* from the tiniest beings, to plants, animals, humans, and the angelic hierarchies beyond.'

'So,' Ali began, 'physical science can tell us about Aristotle's first two causes, MATERIALS and MECHANISMS of nature, but little of the *subjective, consciousness or 'heads'* sides, including the *inner* dimensions of the mineral, vegetable, animal, and human kingdoms—the *Soul* sides of things. Whereas, the world's esoteric traditions[261] can form part of an emerging *spiritual* science, and provide insights into these things?'

'Yes. Religious creation stories along the lines of, "God said, 'Let the Worlds exist,' and they did," reflect our ancient intuition that existence has *Subjective, Spiritual* dimensions. But, they do not describe the history or the physical MECHANICS of Life's workings. This was why, in the 1600s, Sir Francis Bacon, set out, in his **Novum Organum Scientiarum**, an epoch defining program for the development of modern PHYSICAL science.

'One of Bacon's insights was that natural philosophers, as scientists were then called, should focus less on the meanings and purposes of things, their Aristotelian *formal and final* causes, *soulful* as it was, but on a systematic analysis of our physical world's MATERIALS and MECHANISMS.

'They should go as deeply as possible into what things are **made of** and *how they* **work**—focusing on accurate observations, and experiments, rather than on ancient dogmas of how things were *'supposed'* to be, based on old, often untested, theories, as sometimes happened in the middle ages.

'However, once scientists had better understood the materials (made of), and mechanisms (how it works), of the physical worlds, as we now do, they could always return to exploring the *formal and final* causes of things, their *inner meanings* and their *purposes,* a process based partly on deductive logic and, at times, on more *subtle* and more *intuitive* ways of knowing.

'Sir Francis Bacon's program was very successful. It is, in part, thanks to him, and others like him, that the scientific revolutions took off, and we have all the amazing technological benefits we enjoy today.

'Having gained all this practical knowledge, we are now at the point when scientists can start thinking, once again, about formal and final causes, about the *meaning* and *intelligence* of things, and, of *who* and *what* we truly are. We cannot do this by insisting, from the outset that: "Only the physical is real and all ideas of the *spiritual* are fanciful," or that, as MM/MN claims:

> "We can explain everything in purely **physical** or **'natural'** terms, even if the *spiritual* and the *super* natural *are* real."

'If we do that we are no better than the medievals who held things *'had'* to be a certain way to fit their prior beliefs—the Sun *'had'* to orbit the Earth because that's what many then believed. Today it's, "Darwin *'has'* to be right, and materialism *'must'* be true, because is is what we have thought for a long time . . . even if it's false."

'Science, though, is supposed to be an open-minded quest for truth—so,

Was Darwin right? Did Life arise, and evolve, as he said, or not?
Is there *more* to reality than skeptics of the *spiritual* believe, or not?
Is there any evidence for *intelligence* and design in nature, or not?
Can DNA Codes or Living Cells arise by blind chance, or not?
Could the caterpillar transform into the butterfly by 'accident,' or not?
Are reports of *mind-beyond-the-body* real, or not?
Do verifiable *afterlife* communications take place, or not?
Are *psychic* abilities of various kinds possible, and real, or not?
Is there any evidence for them?

'Are these not all interesting questions?'

We continued on our way.

24 Mind Beyond Matter

'Our science is not yet much interested but there are many ways to falsify materialism. One way is to question Darwin's theory—including his idea that Living things may have *'fallen together'* by chance. While purely physical forces can explain some things, like the evolution of landscapes, can they really account for the arising of *Living Beings?*

'Another way is to conduct research into phenomena like *telepathy, psychic* abilities, *spiritual* healing, near death and other *out of body experiences* (NDEs and OBEs). There are powerful, modern, ways to put the evidence for such phenomena to the test.

'A well known contemporary researcher is Rupert Sheldrake. He is a well known biologist willing to study phenomena which do not mesh with materialist interpretations of reality.

'He has written about the fascinating results of some of his researches in *A New Science of Life, Dogs that Know When Their Owners are Coming Home, The Sense of Being Stared At* and *The Science Delusion.*

'Richard Dawkins, on the other hand, is a well known atheist who has devoted a lot of time to dismissing belief in religion or, indeed, in *spiritual* phenomena which materialism cannot explain away. He wrote *The Selfish Gene, The Blind Watchmaker* and *The God Delusion* among others.

'Intriguingly, in 2007, he visited Sheldrake to discuss the latter's research on **unexplained abilities in animals and people**. The interview was intended to form part of a TV series called *Enemies of Reason* which Dawkins was making as a follow up to his critique of religion called *The Root of All Evil.*

'Before *Enemies of Reason* was filmed, the TV company assured Sheldrake—who was cautious about taking part—that the documentary would be "an entirely more balanced affair than *The Root of All Evil* was." And that, "We are very keen for it to be a discussion between two scientists, about scientific modes of enquiry.' So, they agreed to meet. Sheldrake writes at his website [at time of writing here],

> 'I was still not sure what to expect. Was Richard Dawkins going to be *dogmatic,* with a mental firewall that **blocked out any evidence that went against his beliefs**? Or would he be open-minded, and fun to talk to? . . .
>
> 'Richard began by saying that he thought we probably agreed about many things, "But what worries me about you [Rupert] is that **you are prepared to believe almost anything**. Science should be based on the minimum number of beliefs."
>
> [⊛ Unless it's the extraordinary belief that Life could arise by chance. ⊛ Unless it's the extraordinary belief that DNA CODES could write themselves. ⊛ Unless it's the belief that spiders, their silk and webs could arise, and evolve, by pure luck, ⊛ unless it's the belief that a two sex world could arise by pure chance, ⊛ unless it's the belief that…].
>
> 'Sheldrake replies. "I agreed that we had a lot in common, "But what worries me about you [Richard] is that you come across as *dogmatic,* giving people a bad impression of science."

'Dawkins then said to Sheldrake that he'd *like* to believe in telepathy but there wasn't any *evidence* for it. He added that telepathy could not be *possible* because it would "turn the laws of physics upside down." A view which, as we saw earlier, rests on outdated ideas about physics.

'Dawkins also put Carl Sagan's famous point to Sheldrake that "Extraordinary claims require extraordinary evidence." Sheldrake replied:

> "This depends on what you regard as extraordinary. Most people say they have *experienced* telepathy, especially in connection with telephone calls. In that sense, telepathy is *ordinary*. The claim that *most people* are deluded about their own experience is extraordinary. Where is the extraordinary evidence for that?"

'Dawkins replied that people are prone to *wishful* thinking, when it comes to the *super* natural and the *paranormal*. Maybe. This, certainly, is a common claim made by those who, like Dawkins, dismiss all things *spiritual*.

'Yet, these same thinkers always to fail to notice the *astonishing* levels of 'wishful thinking' needed to believe that pure luck could give rise to any *living* thing, to *Cells! and Shells!, to Blood! and Brains!,* let alone to M I N D and C O N S C I O U S N E S S. Sheldrake continues,

> 'We then agreed that [1] **controlled experiments were necessary**. I said that this was why [2] I had actually been doing such experiments, including tests to find out if people really could tell who was calling them on the telephone when the caller was selected at random. [3] **The results were far above the chance level.**
>
> 'The previous week I had sent Richard copies of some of my papers, published in peer-reviewed journals, so that he could look at the data. Richard seemed uneasy and said, **"I don't want to discuss evidence."** "Why not?" I asked. "There isn't time. It's too complicated. And that's not what this program is about." The camera stopped.
>
> 'The Director, Russell Barnes, confirmed that **he too was not interested in evidence**. The film he was making was another Dawkins polemic. I said to Russell, "If you're treating telepathy as an irrational belief, surely evidence about whether it exists or not is essential for the discussion. If telepathy occurs, it's not irrational to believe in it." . . .
>
> 'Richard Dawkins has long proclaimed his conviction that "The paranormal is bunk. Those who try to sell it to us are fakes and charlatans". ... But does his [materialist/atheist] crusade really promote "the public understanding of science," of which he is [at the time,] the professor at Oxford? Should science be a vehicle of prejudice, a kind of fundamentalist belief-system? Or should it be a method of enquiry into the unknown? '[262]

'That's interesting, Ollie, two scientists with very different attitudes...'

'Yes, the two very different approaches to inquiry we have been discussing. The first assumes, as a matter of methodology (MM/MN) or philosophy (PM/PN),[263] that reality is *only* PHYSICAL, that it must be studied as such, and all things, including *Life, Mind, Emotions, Consciousness* **must** have purely physical explanations. To suggest anything different, is, as *wikipedia* puts it, a 'religious' argument—and therefore not allowed to be part of science today—even, if it is actually true!

'The second, Sheldrake's approach, is to remain open to a wider range of possibilities, including the classical hypothesis that there may be more to *Life* than (mindless) **matter** + (mindless) **laws** + (mindless) **chance**.

'Another scientist, taking a broader view, is Harvard academic, Dr Gary Schwartz, Ph.D, author of ***The Afterlife Experiments****: **Breakthrough Scientific Evidence of Life After Death***. Dr Schwartz "asked some of the most prominent [spiritualist] mediums in America to become part of a series of experiments to prove, or disprove, the existence of an afterlife."

'His investigations led to "a breakthrough scientific achievement: **contact with the beyond under controlled laboratory conditions**."

> 'In **stringently monitored** experiments, leading mediums attempted to contact dead friends and relatives of 'sitters' who were **masked from view** and never spoke, depriving the mediums of any cues. **The messages that came through stunned sitters and researchers alike.** There are some extraordinary and uncanny revelations . . . Dr. Schwartz was forced by the over-whelmingly positive data to abandon his skepticism.' [264]

'After the experiments, Dr. Schwartz wrote,

> 'I can no longer ignore the data [on life after death] and dismiss the words [coming through mediums]. They are as real as the sun, the trees, and our television sets, which seem to pull pictures out of the air.'

'But aren't these things more in the realms of *religion* than science?'

'No. While it *is* true that the religious have their own convictions about these things, others, not committed in that way, can conduct their own, rigorous research, in a *non-religious* way. The Australian afterlife researcher, Victor Zammit, makes this point. He writes,

235

"The [plentiful] objective [modern] evidence for the after life has nothing to do with religion or personal belief." [265]

'So, why is there so little mainstream interest in this kind of research?'
'Because, Ali, as Chris Carter says, it is the,

"implicit **equation of materialism with science,** that explains the widespread practice of ignoring and dismissing the objectionable evidence as somehow 'unscientific.'

"Materialism is upheld as **an incontestable dogma** on which, it is thought, rests the entire edifice of science. But the difference between science and ideology is not that they are based on different dogmas; rather, it is that **scientific beliefs are not held as dogmas,** but are open to testing and hence possible rejection.

'Science cannot be an objective process of discovery if it is wedded to a metaphysical belief [like materialism] that is accepted without question and that leads to the exclusion of certain lines of evidence on the grounds that these lines of evidence contradict the metaphysical belief.' [266]

'Can you put that more simply?'

'His point is that science cannot be *objective* if it assumes, from the start, that **materialism** is true—when this is the very question. *Is* materialism true? Has it been *shown* to be? Can *Life, DNA CODES, Cells, Shells, Worms, Fish, Birds, Dogs, Humans,* with very clever *Minds,* really emerge, for no reason at all, from MINDLESS matter, by 'dumb' luck?

'Another of the modern challenges to materialism is the extensive evidence for *mind beyond the body.* The many—non religious—accounts of near death and other *out of body* experiences documented by organizations like the International Association for Near Death Studies and the Monroe Institute in Virginia. There are, also, many good books on these topics. Raymond Moody's *Life After Life,* Margot Grey's *Return From Death,* the Fenwicks *The Truth in the Light: An Investigation of over 300 Near Death Experiences,* and many, many others.

'In *Science and the Near Death Experience: How Consciousness Survives Death,* Chris Carter examines ancient and modern accounts of NDEs from around the world, including the West, India and China.

'A reviewer writes,

> '*Science and the Near-Death Experience* ... [is] "the best book on NDEs in years." Its strength is (among others) that it takes all 'skeptical' explanations seriously, examines them thoroughly and demonstrates why they all fail. The necessary conclusion is that NDE/OBEs **are, in fact, what they always seemed to be**, and what all experiencers hold them to be: Experiences of a mind, which has left its body.
>
> 'Perhaps you will insist that "science" cannot accept this conclusion, because it must adhere to a materialistic monism?
>
> 'Well, **what is the task of science? Is it to explain phenomena** or to *explain them away*, following a pre-set ontology? Is the ontological basis for science testable or not? If it is not, that basis is in effect a dogma. And science is no longer science.' [267]

'What does, "**Is the ontological basis for science testable or not**?" mean?

'He means that if scientists base their ideas on a particular set of *beliefs* about the *basic nature* of *reality,* which is what *ontology* means, then, for their claims to be *scientific,* their beliefs must be *testable*. Otherwise they're just *opinion* and *ideology*.

'Here are some conflicting ontological bases which can be tested.

(1) The Cosmos is MATERIAL alone. No, it can be *shown* to be more.
(2) *Mind,* even if it were a principle in its own right, could *not* influence MATTER. No, it has been shown that mind *can* influence matter.
(3) Scientists can only detect *'unintelligent design'* in Nature. Not true. They are very well equipped to detect both unintelligent design (random events) and intelligent design (meaningful patterns).
(4) Life is just an accident. No. It's demonstrably impossible.
(5) Mind cannot exist beyond the body-brain. Yes, it *can*—as has been *repeatedly* shown, with many NDE, OBE and afterlife studies.

(6) Life after death is not possible. False. It is possible because there is a great deal of good quality, modern evidence for it.

'The answers to these kinds of *ontological* questions will determine the kinds of inquiries scientists engage in. If they have a prior commitment to materialism, if they think, from the off, that reality is PHYSICAL alone, not only will they not be interested in investigating any of its *super* natural or *multi-dimensional* elements, they may actively seek to discredit any claims about them, as Richard Dawkins has sought to do.

'On the other hand, if, like Rupert Sheldrake and Alfred Russell Wallace, they think *mind* may be a *principle* in its own right, they will be more likely to conduct such research.

'In *To Heaven and Back: A Doctor's Extraordinary Account of Her Death, Heaven, Angels, and Life Again: A True Story,* scientist and surgeon, Dr Mary Neal, describes her near-death experience. In 1999, in the Los Rios region of southern Chile, while descending a waterfall, her kayak became pinned at the bottom. She experienced death by drowning, and left her body for a period, before being revived a short while later. In a subsequent interview she reported:

> 'I was acutely aware of everything that was happening. I knew that my efforts to exit the boat were not working, that I was out of air, and that I was too far from the riverbank for anyone to reach me. I knew that I would probably die. Having grown up with a fear of drowning, I was surprised to find my transition from life to death was seamless, peaceful, and beautiful. I felt quite wonderful.
>
> 'Before my near-death-experience, I believed in God and took my kids to Sunday school but was not particularly religious. Like many accomplished young adults, I felt like I was in control of my life and my future. Although I tried to be a "good" and "moral" person, my faith was not integrated into my daily life and the demands of work and family left little time to think about spirituality.' [268]

'Neal describes herself as "a scientist by training, a skeptic by nature, and a very concrete, rational thinker." Yet, as for many other NDErs, the *spiritual* impact of her experience was major. Prior to this experience, she was not religious. After, spirituality became a very important part of her life.

'When asked why she thought she'd come back, she replied, like many other NDErs, that, having seen the 'other side,' she didn't *want* to return, but felt she had little choice, that she needed to share her experience with others, and to support them in *their* faith. When asked what she'd like people to know about the afterlife she says, simply,

> 'God's unconditional love for each of us is intense, complete ... Before we return to Heaven, our real home, we have an incredible opportunity on Earth to face challenges that will help us learn, grow and to become more Christ-like . . . Our time is so short that we need to be about God's business every day.' [269]

'Here is another remarkable NDE quoted by Dr. Bradley Nelson in his best-selling book on emotional healing, ***The Emotion Code***. Anne Horne from Seattle wrote to him as follows:

> 'I originally came into contact with your [healing] work when a practitioner... used your techniques to balance my body... [and, at the end of the session, she] said, 'Oh, let's see if you have a Heart Wall...' She explained how emotions can put up a wall between yourself and others around your heart. [And how] she would run a magnet down a person's back to release those emotions and open their hearts.
>
> 'I couldn't believe what I was hearing. It was like an electric bolt went through me. Suddenly, an event that happened to me 25 years ago made sense... When I was 23... I had one of those NDEs . . .
>
> 'It was a very important experience for me. I left my body and had a life review... I was going back, home, and on my way, there was a light, a tunnel. I felt like I was being pulled by my heart toward a wonderful place. In that moment, I was encompassed by all this *innate intelligence and tremendous love*. I just wanted to go home. It was fantastic.

'Anne was then told, to her disappointment, that it was 'not her time' but she was shown something in her future, on Earth, where she saw people,

> 'doing a specific training that was very unusual. There would be 2 or 3 people together, with one person lying on a table, or standing up, and another person who was rolling something down the other person's back.
>
> I knew this was in my future, that I was one of these people. I could feel the sense of urgency that they were feeling. It was like a numbers game; we had to treat as many people as possible. We were really in a rush, really hurrying. It was very, very, very vital. I couldn't quite understand what was going on, [so I asked him] 'what are we doing?' 'You are opening people's hearts,' he said. 'Not in a physical way. You're removing all blocks from their hearts so that they can give love and receive love from here.'
>
> At that moment, the people doing this work became consciously aware of each other. It wasn't something planned . . . It just happened. We became conscious of each other. And at that moment, the meaning of this work became clear to me.
>
> Suddenly a flood of energy was sent to the earth from where I was, above the earth. It looked like a white bolt of energy that came in through... our hearts into the world. We were there opening people's hearts so that they could be anchors for this *divine energy* to come into this world. Within three seconds, the world was completely transformed by this energy. This *light* went into every crack and crevice, everywhere, and there was no darkness in this world, ever again. The next thing that happened was, the doctors resuscitated me and I was brought back to life.'
>
> 'During my near-death experience, I said to the man who met me, 'But there are only thousands of us.' And he replied, 'Millions will hear, but only thousands will remember. And we only need thousands. Thank you for giving me a way to fulfill my mission.' [270]

'Following her NDE,[271] Anne was inspired to pursue various therapeutic trainings and, eventually, to become a practitioner of the *Emotion Code*, when she understood what she had seen in her experience.

'In ***Dying to Be Me***: *My Journey from Cancer, to Near Death, to True Healing,* Anita Moorjani describes her four year battle with cancer during which her body eventually became horribly emaciated and full of tumors.

'She was confined to a wheel chair, unable to breathe without extra oxygen and, eventually, fell into a coma when, it was believed, she would soon die. Anita describes how, at the hospital, she found herself outside of her comatose body, in an expanded state of awareness with 360 degree perception. She *understood* the oncologist, beyond the range of normal hearing, 40 feet from her bed, explaining to her husband that, although his wife's heart was still beating 'she's not really in there' and it's 'too late to save her.' She observed the medics working on her body but noticed she no longer felt any attachment to it. In fact, she felt liberated, with every 'pain, ache, sadness and sorrow' gone. She had never felt so good before.

'She noticed feeling *weightless,* with a sense of knowing that she could be anywhere at any time, and that felt normal to her, as though returning to a more accurate way of perceiving reality. She heard the doctor saying,

"There's nothing we can do for your wife, Mr. Moorjani. Her organs have already shut down. ... She won't even make it through the night." She found herself longing to tell her husband that she felt ok, that *all* was ok. As her awareness continued to expand, she realized that her brother was on a plane, hoping to say goodbye. She wanted so much to tell him she was fine. In fact, she had never felt better. Next, it came to her, with a sense of surprise, that the most *basic* nature of the cosmos is *love.*

'She noticed herself wondering, why could she see and understand so much? Who was giving her this information? Was it God? Buddha? Jesus? To her joy and delight, she understood at last that the divine can be *known, directly, as a state of **Being**,* that she *was* that state of *Being* and that,

> "I'm not who I'd always thought I was: Here I am without my body, race, culture, religion, or beliefs . . . Yet I continue to exist! Then what am I? Who am I? I certainly don't feel reduced or smaller in any way. On the contrary, I haven't ever been this huge, this powerful, or this all-encompassing. Wow, I've never, ever felt this way! . . . I felt eternal, as if I'd always existed and always

would without beginning or end. I was filled with the knowledge that I was simply magnificent!" [272]

'She found herself wondering why she had always been *so hard* on herself? Because, she now *knew, deeply,* that she was loved, *deeply, unconditionally.* Why, then, her constant need to achieve and win approval?

'Why had she not loved herself unconditionally? Why did she not have this understanding when *'in her body?'* Why had she not realized that we're not supposed to be so *hard* on ourselves? She found herself immersed in a sea of unconditional love and acceptance.

> "I became aware that we're all connected. This was not only every person and living creature, but the interwoven unification felt as though it were expanding outward to include everything in the universe – every human, animal, plant, insect, mountain, sea, inanimate object, and the cosmos. **I realized that the entire universe is *alive* and infused with consciousness**, encompassing all of life and nature. Everything belongs to an infinite Whole. I was intricately, inseparably enmeshed with all of life. We're all facets of that unity – we are all One, and each of us has an effect on the collective Whole. . ." [273]

'Very much like what the world's first peoples have always said...'

'True... Anita, then faced a choice. She experienced her father, already passed on, communicating to her that she could either continue into the spiritual worlds, and the cause of death would be organ failure, or, she could return to her body and recover. Like many other NDErs, she found the thought of returning *very uninviting.* But, if she did, she would be well.

'One of the really inspiring things about Anita's NDE is her extremely rapid subsequent recovery—it's medically extraordinary. You should read it.'

'Ok, that's another intriguing account. But these stories, truly inspiring as they are, are *anecdotes*. They have no scientific merit, do they?'

'Well, there *are* patterns here, especially if you review the NDE literature in depth, as Chris Carter does in ***Science and the Near Death***

Experience, and Raymond Moody did in *Life After Life.* Is it fair to dismiss people's experiences just because we don't have personal access to them?'

'But are people's *private* experiences, of any public relevance? And, what relevance do they have to our conversations about evolution?'

'I mention them, Ali, because, currently, our science treats natural history as a totally MATERIALISTIC process—no *spiritual* input allowed. Regrettably, western science has rejected Cartesian dualism—*and* all ideas of *Spirit, Life Forces, and Soul*. So, in the west today, to be called a *'vitalist'* or a *'spiritualist'* is to be seen as unscientific and disreputable.

'On the other hand, the modern, **non-religious** evidence from NDE's, OBEs and other psychic phenomena *falsifies* materialism. It supports the classical hypothesis that *no* wholly materialistic theory of Life, be it Darwin's, or anyone else's, will be adequate to give a *full* account of reality.

'As to the public relevance of people's *private* psychic experiences there is now such a large body of evidence and research relating to NDEs and other kinds of OBEs that it is blinkered and unscientific to ignore it.

> 'A college professor ... told of dying on the operating table and then instantaneously finding himself walking down the gentle slope of a bright greensward. No one else was there. Nothing else happened. Yet to hear him speak of it, it was as if the greatest of miracles had happened to him, and **he now knew, he absolutely knew, that there was life after death**. This brief incident completely transformed his life. Even today he lights up describing the incredible aliveness of the green grass he once walked upon on The Other Side of death.'
>
> 'Atwater, continues, 'I have spoken with many people who have described 'the living dark' that [first] greeted them with words like *'soft velvety blackness'* and *'warm inviting blanket'*. ... these people felt awed by the wonderment of a blackness that appeared intelligent, emoted feelings and instilled in the experiencer a sense of peace and acceptance. . . .
>
> '[Others] encountered light . . . They claim that the radiant brilliance of this special light does not blind or burn; **it simply**

accepts, embraces and loves. A sense of worthiness can remain afterwards, changing forever how the experiencer regards him or herself.' [274]

'NDEs are personal experiences but they are of interest to all of us.'
'Okay...'
'Because they show us that *A W A R E N E S S* and *M I N D* are not merely aimless *epiphenomena* of *'purposeless'* bodies, but, rather, are transmitted by them, just as TVs transmit their *shows* from elsewhere.

'Often, those who have had these kinds of experiences are greatly *changed* by them. They're more *loving* and more at *peace,* and they no longer fear death. Surely it's a very good thing for us, as a world culture, to become more aware of these things.'

'Ok, it might reduce our fear of death. But are these people really dead?'
'Yes, in some cases they are judged to be clinically dead.'[275]
'But isn't there a difference between OBJECTIVE and *subjective, or inner,* experiences?'
'It depends on what you mean by objective. A tree is not made OBJECTIVE just because we can all see it.

'We all experience the objective tree *subjectively*. *All* experience is *Subjective*. Only *Subjects*, **only knowing Beings**, capable of *knowing Being, capable of* **knowing Existence***,* can experience. There can *be* no existence, no *Being, of any kind,* without *Subjects, or Beings, or Experiencers* to know it.

'Currently, our science endorses our *collective* subjective experiences of the tree and it calls them OBJECTIVE. But, strangely, it dismisses the truth of numerous individual experiences, like NDEs and OBEs, even when they share many commonalities and can sometimes be validated. [276] No one's *subjective* experience, even of a tree, which we can all see, is fully accessible to anyone else. Yet, in normal reality, we take ***account*** of other people's experiences. For example, we often rely on just one or two witnesses to convict someone of a crime. Yet, even when ***many*** people, all over the world, have testified to various *mystical, psychic,* and other *spiritual* experiences, even though their experiences share many commonalities, and can, at times, be validated and cross-checked, they are often dismissed.

'Why? Because, as Chris Carter writes, our science, currently so captured by materialism, "cannot accommodate corroborated reports" that human consciousness can operate outside of a physical body.[277]

'This is why any evidence which challenges **materialism**, such as NDEs, OBEs and messages from those passed on[278] is dismissed or ridiculed.

'But why are they so dogmatic about this . . ?' Wondered Alisha.

'Because, they mistakenly *equate science*—which meant to be an open minded quest to find out *what is true*—with `materialism`, which is a `belief system` or `ideology`, which may or may not be true.

'In fact, "When many scientists and philosophers are confronted with the evidence, their reaction is often anything *but* rational." Carter quotes philosopher of science, Neil Grossman, who reported that he,

> 'was devouring everything on the near-death experience I could get my hands on, and eager to share what I was discovering with colleagues. It was **unbelievable to me how dismissive they were** of the evidence. "Drug-induced hallucinations," "last gasp of a dying brain," and "people see what they want to see" were some of the commonly used phrases. One conversation in particular caused me to see more clearly the **fundamental irrationality of academics** with respect to the evidence against materialism.
>
> 'I asked, "What about people who accurately report the details of their operation?"
>
> '"Oh," came the reply, "they probably just subconsciously heard the conversation in the operating room, and their brain subconsciously transposed the audio information into a visual format."
>
> '"Well," I responded, "what about the cases where people report *veridical* perception of events remote from their body?"
>
> '"Oh, that's just a coincidence or a lucky guess."
>
> 'Exasperated, I asked, "What will it take, short of having a near-death experience yourself, to convince you that it's real?"
>
> 'Very nonchalantly, without batting an eye, the response was: "Even if I were to have a near-death experience myself, I would

conclude that I was hallucinating, rather than believe that my mind can exist independently of my brain."' [279]

'For inflexible people like this, **materialism** is no longer just another hypothesis about reality open to "being proved false." No. It's "an article of *faith* that *"must"* be true, regardless of evidence to the contrary." [280]

'In *Consciousness Beyond Life, The Science of the Near Death Experience,* cardiologist Pim van Lommel describes how, **over twenty years, he interviewed hundreds of heart patients** who had died, some for five minutes or longer, before being resuscitated. Of these, some reported ongoing experience after medical equipment reported them dead.

'**Some recalled witnessing the actions of hospital staff from out of body perspectives**. Do we have to assume these experiences *cannot* be real—just because they falsify materialism? Just because some of us—ignoring Life's miracles, staring us endlessly in the face—cannot conceive that reality is *multi-dimensional, not* mono-dimensional, and *Soul* and *Spirit* are real!

'Here's another interesting account. Dr Larry Dossey describes how, during a routine gallbladder operation, a woman experienced cardiac arrest and died. After she was resuscitated, **she was able to describe some of the events taking place around her, while out-of-body, in vivid detail**. She noticed that the anesthesiologist was wearing mismatched socks. What makes this account especially interesting is that Sarah had been blind since birth.[281]

'There is, now, **decades of good quality, *non-religious* research** into NDEs and OBEs of various kinds, into communications from those passed on and into the modern evidence for reincarnation.[282] Research which shows us that there are d i m e n s i o n s *beyond* the PHYSICAL where those passed on continue to live and to evolve, including possibilities for reincarnation.[283]

'Typically, though, these findings, of researchers at bodies like the British and American Societies for Psychical Research, the International Association for Near Death Studies, the Monroe Institute, the Afterlife experiments of Dr Schwartz, the medical researches of neuropsychiatrist Dr Peter Fenwick, and cardiologist Pim van Lommel, demonstrating that out of body and afterlife phenomena are real, have been ignored or ridiculed.

'Why? Why aren't more people interested in this modern research?'

'Because it can be difficult for us to admit that we may have been mistaken? Even if, by doing so, existence becomes so much broader, so much more meaningful, and so much more interesting to explore…

'After all, in a *multi-dimensional* cosmos of intelligence and meaning, there are other dimensions and planes of existence to investigate—in addition to the ones we are presently familiar with. Just as we explore the physical world, using our physical bodies, so, some say, it is possible, even while we are physically alive, to enter the *spiritual* dimensions, by using our *subtle, non-physical* bodies, as some shamans and yogis are able to do.

'In *Less Incomplete: A Guide to Experiencing the Human Condition Beyond the Physical Body,* Sandie Gustus asks:

> 'How many of us have had an experience that suggests a deeper, unseen reality...? ...déjà vu, intuition, synchronicity, premonition or telepathy; felt an instant sense of recognition or familiarity with a complete stranger ...
>
> 'Despite this, most people have no direct experience that we live in a *multidimensional* environment that extends far beyond the boundaries of our physical world, and that we are, in fact, much more than just our physical bodies. Fortunately, there is one phenomenon that is natural to all humans that allows us to verify for ourselves, from first-hand experience, that we are capable of acting entirely independently of the physical body in a nonphysical dimension ... The **out of body experience** (OBE).
>
> 'Anyone who has had a fully lucid OBE, and I count myself among them, will tell you that **if you can be lucid outside the body, you will find that all your mental faculties are fully functioning**, that you can make decisions, exercise your free will, access your memory, think with a level of clarity that sometimes exceeds your usual capacity, and even capture information from the physical dimension that can later be corroborated . . . and that **to experience all this provides you with irrefutable evidence that the physical body is merely a temporary 'house'** through which your consciousness (i.e. your soul or spirit) manifests in the physical world.'[284] We walked on, the shadows starting to lengthen.

25 Proof of Heaven

'When particle physicists report their findings to us, Ali, we believe what they say. Yet most of us have no *direct way* of testing their claims any more than we can test the claims of a shaman, a high-powered yogi or a psychic.'

'Ok, but isn't it a case of, 'If you studied these matters, you also would be able to access the information, unlike spiritual knowledge, which you have to take on trust, as take it or leave it *revelation?*'

'People do sometimes say that, but it's not true.'

'Ok. . .'

'Because, just as some of us can train to be particle physicists, we can train in extra-sensory perception and, as Sandie Gustus says, explore *other dimensions* of reality. In fact, as the theosophists, Besant and Leadbeater demonstrated, it is even possible to study *particle physics* by means of ESP.

> 'Analyzing twenty-two diagrams of the hundred or so chemical atoms described in *Occult Chemistry*... Phillips found it hard to avoid the conclusion that "Besant and Leadbeater did truly observe quarks using ESP some **70 years before physicists proposed their existence**". [285]

'While most of us are not interested in learning such techniques, any more than we are interested in learning particle physics, in theory we could.

'In, and ever since, the 19th century, perhaps many people do not realize, there has been a *huge* amount of research showing that psychics and spiritualists can provide good evidence for ESP and for Life after death.

'In *Science and the After Life Experience: Evidence for the Immortality of Consciousness,* Chris Carter explores the *extensive* modern evidence for Life after death. He discusses **125 years of studies** by independent researchers and the British and American Societies for Psychical Research which rule out hoaxes and hallucinations. Studies which show that the afterlife is real. One very long-running study was called the *cross-correspondences*. It was initiated by Frederick Myers, a Cambridge classics scholar, a gifted man, whose transmission theory of the mind–brain relationship was taken up and developed by his friend, the author Williams James.[286]

'Myers was one of the founders of the British Society for Psychical Research (BSPR) and he wanted to show, **beyond reasonable doubt**, that information transmitted through spiritualist mediums was not merely the result of a psychic or medium's ability to read the minds of surviving relations—an idea called *super ESP*—but evidence for life after death.

'He proposed that, after he had passed on, he would attempt to send a series of telepathic messages to *different* mediums in different parts of the world, messages which, on their own, would be meaningless but, *if* put together, would make sense. He and his colleagues at the BSPR believed this would provide strong evidence for survival, rather than just super ESP; although super ESP would be very interesting in itself.

'After Myers passed on, in 1901, more than a dozen mediums, in different countries, began receiving a series of incomplete scripts which they channelled through automatic writing. The pieces were signed 'Frederick Myers.' The scripts contained obscure material from—and references to—the classics which made no sense on their own. The messages did, however, ask the mediums to contact a central address, in London, where the scripts were assembled and, then, made sense.

'Later, Myer's colleagues at the BSPR, Professor Henry Sidgwick and Edmund Gurney, also transmitted scripts after passing on. More than 3000 scripts were transmitted over 30 years, some more than 40 pages long.

'In total, 12,000 pages, in 24 volumes. Some time later, Colin Brookes–Smith patiently studied this material and stated, in the *Journal of the BSPR,* that after-death survival should now be regarded as sufficiently well-established to be beyond denial by any reasonable person.[287]

'In *Science and the After Life Experience* Chris Carter investigates a variety of historic and contemporary accounts of past-life memories, supersensory visits from those passed on, telepathic communications via mediums and automatic writing, as well as the methods used to confirm these experiences.

'Notably, in examining the evidence for these phenomena, he carefully explores **alternative** explanations. Only when these have been ruled out does he conclude that the phenomena really are what they seem to be.

'Dr. Pim van Lommel, author of *Consciousness Beyond Life,* writes: "The evidence in favor of an afterlife is vast and varied. . . .we do indeed,

> "have strongly *repeatable* evidence for the continuity of consciousness after physical death . . . What all these cases show is that ... **human consciousness can exist independently of a functioning brain**. When one has read the overwhelming [modern] evidence as described in this [Chris Carter's] excellent book, it seems quite impossible not to be convinced that there should be some form of life after death. Any continuing **opposition to the evidence is based on nothing more than willful ignorance or ideology**." [288]

'You see, Ali, it is not good quality evidence for *super* natural phenomena which is lacking, but the equation of science with the `materialist ideology` which prevents its mainstream acceptance. One thing which amuses me, though, is how many self-styled *'skeptics'* of the *super* natural demand ever more proof for it, even though the evidence, which they call for, no matter how good, is never good enough for them.

'Yet, these same, self-styled *'skeptics'* are quite untroubled by the lack of proof for Darwin's theory, let alone for the claim that mere chance could be the cause of their own incredibly complex bodies and their own clever minds.'

'Dr. Larry Dossey, author of *The Power of Premonitions* and *Healing Words,* is more scathing than Dr. Lommel, he writes:

> 'Chris Carter's *Science and the Afterlife Experience* ... is a **withering rebuttal of the perennial, timeworn, anemic arguments of [so called] skeptics**. This book is extraordinarily important, for, as Jung said, '**The decisive question for man is: Is he related to something infinite or not**? That is the telling question of his life.'

'This brilliant book is an antidote to the fear of death and annihilation. It will help any reader find greater meaning, hope, and fulfillment in life.' [289]

'Earlier than Carter, some very distinguished, yet initially skeptical, scientists also investigated *afterlife* communications, often with a view to debunking them. Yet, as they continued to inquire, they concluded that *multi-dimensionality,* that *Life* and *Mind* beyond the body were facts.

'Sir William Barrett F.R.S. (1844–1925), Professor of physics at the Royal College of Science in Dublin, reported that he was, "absolutely convinced of the fact that those who once lived on earth can and do communicate with us. It is **hardly possible to convey. . . the strength and cumulative force of the evidence**."

'Sir Oliver Lodge F.R.S. (1851–1940), famous for his pioneering work in electricity and radio, and for developing the spark plug, concluded that "**survival is scientifically proved by scientific investigation**."

'Professor Camille Flammarion (1842–1925), founder of the French Astronomical Society, who investigated psychic phenomena for decades, came to the same conclusion.

'Professor James J. Mapes (1806–1866), an expert in chemistry, set out in the 1850s to *'rescue'* his friends involved with the modern spiritualist mediumship movement which had become very popular at the time. Yet, after investigating many mediums he changed his views and stated,

> "The manifestations . . . are so conclusive [that they establish]: First, that there is a future state of existence, which is **but a continuation of our present state of being**. . . Second, that the great aim of nature, as shown through a great variety of spiritual existences is progression, extending beyond the limits of this mundane sphere . . ."

'These eminent scientists and thinkers, Ali, were all, sober, measured people, often highly skeptical when they began their investigations.

'It was only after a lot of research that they accepted the evidence. Many of them were extremely practical people, who had made major, sometimes world-changing discoveries in other fields. They were free thinkers who braved opposition not only from the materialists but also from some of the traditional religious authorities.

'Why would religious authorities oppose this kind of evidence, for NDEs, OBEs and for Life after death? Surely the modern research evidence would strengthen their own ancient claims for a multi-dimensional cosmos of intelligence and meaning?'

'Because the modern evidence for the *afterlife* can be challenging to long held religious views. For example, outdated ideas about 'heaven and hell. The evidence, from NDEs, OBEs, and telepathic communications from those passed on, being that, in certain important respects, Life after death is not so different from the life we already know.

'We still experience shapes and forms, sentiency and feelings, sights and sounds, textures and tastes, at different vibrational frequencies from here. Then, it is our own qualities of *mind,* and *Soul,* which govern *'where'* we find ourselves, in the after-life realms, not an irrational scheme of eternal punishment, or a boring, over-pious heaven.

'Something else, from modern psychical research, which conflicts with traditional religious teachings, is that most messages, from the afterlife, tell us it's not so much the *creed* you adhered to, or whether you had one at all, but the quality of your *character,* how you *acted* towards your fellow beings, while on earth, which is the measure of your *spiritual progress.*

'Did you seek the good, the beautiful and the true? Were kindness and the golden rule your guides?' Not, 'Did you believe precisely everything you were told to believe, or to do, by your particular religious tradition?''

'If only we knew more about all this . . .' said Alisha, wistfully.

'Well, we are steadily learning more. Right now, our mainstream science does not accept the powerful, *modern* evidences for these things, nor is it interested in the idea that we exist within a *meaningful, multi-level,* cosmic system. No, the dogma is: "Life is just an accident, which evolved by accident, and Darwin proved it! To suggest or argue otherwise is a 'religious argument.'" Remember *wikipedia's* irrational attack on intelligent design?

'But what could be more interesting than whether there is anything beyond our apparently **so solid** world, consisting almost entirely of S . . . . . p . . . . a . . . . c . . . . e ? 9 9 . 9 . . . 9 . . . 9 . . . 9 . . . 9 . . . 9 . . . 9 . . . 9 . . . 9 . . . 9 . . . 9 . . . 9 . . .% of it, mused Alisha.

'I agree, Ali. We like to think of ourselves as an inquiring, scientific, and open minded society, yet, strangely, so many of those who think themselves *intellectual or scientific* are remarkably *incurious* about these things.'

'Why though . . ?'

'Partly, because too many of us just follow the prevailing views. Partly, a fear of the unknown, including of the *super* natural. Partly, it is the ethos of our time, where too many now conflate **rationality** with **materialism,** and any talk of the *spiritual* with superstition. It doesn't work, it doesn't make sense, but, as *wikipedia* says, we are to believe in '**unintelligent design!**''

'It is also because many self-styled *'skeptics'* of the *super* natural are closed minded. The evidence for all kinds of *spiritual* phenomena, collected over decades, doesn't match their prejudices, so they either allege trickery, or demand ever *'more proof,'* even as they refuse to properly study the very evidence and the very proof which they, themselves, keep calling for.

'Another tactic, they are prone to, is to move the goalposts for the 'proof' they desire, or to raise the bar of proof ever higher, even as they happily go along with the scientifically impossible—and the mathematically absurd—idea that chemicals in "a warm little pond" could, by pure chance, become Life, and, eventually human beings like you and me!

'How much nonsense, Ollie, the rest of us are urged to believe, just because some of us are disillusioned with the traditional religions, because the creation stories in the scriptures turned out to be symbolic, not science texts, and because we cannot see Souls coming and going in our telescopes!'

'Ah, Ali, I could not agree more! As this reviewer of Robert McLuhan's ***Randi's Prize: What Skeptics Say About the Paranormal, Why They Are Wrong and Why It Matters*** notes,

> '[Psychic phenomena] have been studied in painstaking detail for 150 years by highly qualified scientists, including, in recent

decades, **some of the most carefully executed scientific experiments ever conducted** with multi-layered experimental controls that put other fields of science to shame.

'Because researchers in this field are under unrelenting, often vicious assault, **they control even against [even] absurdly improbable and unrealistic forms of cheating** and fraud among other things, problems that most scientists don't have to think about at all . . . **Statistically and taken as a vast body of work, their results are rock solid** . . .

'The scientific facts are in; they're well-proven and extensively documented – many tens of thousands of pages of detailed studies.[290] **The demand for [ever] more proof is simply a ploy [not to awaken, not to realize, we may have been mistaken].**' [291]

We continued on our way, hills darkening, birds, rabbits, cattle, sheep, settling for the night, river quietly flowing.

26 The Great Mystery

'Our amazing world and the velvety infinity of starry space beyond, all come from a *"what-ever-it-is"* which *"just Is,"* be it mindless or be it intelligent. If intelligent, it does not *force* itself on us. We can *believe,* and *be,* whatever we like—mystical, religious, agnostic, atheist, materialist—whether or not our ideas accurately accord with reality, with **what really Is.**'

'You mean free will?'

'Yes. Our gift and our challenge. But, some say, if we use our free will to still our busy minds, to meditate, to visualize, or to pray, we may find that *the subtle, ever-beyond-full-understanding, yet still somewhat-knowable final Source, and Uncaused Cause, of All That Is,* from whose *Shimmering Living Fields* we emerge into this, our temporary physical reality, quietly whispers to us, in *subtle* ways, to tell us *who* and *what* we really are.

'Not, after all, orphan children of a heartless existence which doesn't care for us, because it cannot care, but precious offspring of an *amazing, self-existing, Intelligence-Love-Life-Beingness* which is not only intelligent but loving. Not, Mindless, Unintelligent, and Accidental, or MUA, but an Amazing, Ultimate Miracle, of *intelligent, loving Being,* AUM.

'Mighty A-U-M,[292] mysterious trinity of mind-intelligence-creativity, love-feeling-wisdom, will-purpose-action, reflected in us human microcosms, in Head, Heart, Belly and Limbs. The mysterious creation, which, as *Christian* tradition tells us, is not just *impersonal (It)* but also *personal (It/Is/I(AM)I/S/He/We/You/Are),* ISHWYA, in whose *living image* we, as *Beings of Consciousness, Love, Purpose and Creativity* are also made.

'You are going *mystical* on me...' smiled Alisha softly.

'Maybe, my beloved friend… but existence *is mysterious,* unfathomably so. A stupendous, *impossible, miracle,* emerging from a stupendous, *impossible* mystery. Because, how can there be *anything* at all, let alone a *cause* of anything at all? Isn't it easier to imagine there being *no-thing at all* rather than *some-thing at all?* How on earth, or heaven, or space, is it *possible* for there to be **any Thing-ness or Being-ness** of any kind, even one atom, let alone the **Causing or the Knowing** of it? No one can say. Yet, due to the *infinitely, unfathomably, mysterious, ultimate Source of it All,* which, so astonishingly, *just Is,* the seemingly *impossible* has become *possible.*

'We *know* we *exist* because we are *aware.* Rather than saying, *"I think,* therefore I am," Descartes could better have said, "I am *Aware,* therefore I am." Thinking, feeling, seeing, touching, all need the *Awareness, of the Subject, the Self, the inner Experiencer, experiencing, witnessing Being.*

'We cannot *see* our inner *Self,* our *Awareness,* our *Mind.* But, we can, most definitely, see the clever *effects* of our minds. Yet, when we look at the magical, astonishingly complex, *Living World* all around, we forget this most obvious fact, and we start to claim, in the name of science, that:

> "There is *no inner intelligence or 'purpose,'* to any of it. *Spiritual, Mind* cannot be a principle in its own right, which *'just Is.'* No, only apparently dull matter and its mindless laws can *'just be.'*

'This dull, pedestrian approach can give us no real ideas as to what *Awareness and Love, Intelligence and Curiosity, Feelings and Creativity* truly are. It tells us nothing of the real *essences* or the true *Life* and *Soul* of things.

'Yet, I do sometimes wonder, Ali, whether anyone, no matter what they may *say, really* believes, in their heart of hearts, in the name of science, that amazing, **DNA CODED** Cells, Butterflies and Oak Trees, Eagles and Whales, Apes and People, are all just *chemical accidents? Really? Honestly?*

'If all these beautiful things are *not* to be understood as amazing expressions of *Intelligence, Beauty and Soul* made visible, I think we no longer know what these words *mean,* and important parts of our sciences are no longer instruments for understanding, but irrational vehicles for peculiar distortions which, far from enlightening us, darken our understanding of *who* and *what* we truly are, and, indeed, of the *amazing, living,* nature all around us. When some say, as Darwin did, that life's hardships and nature's harshnesses make them doubt whether there is any *positive Higher Power,* which **Just Is**, *to give rise to All That Is,* I can understand.'

'So how do you answer their doubts?'

'Well, whether something, like a virus or a poisonous spider, [most spiders are harmless to us], has an *intelligent* source, is quite apart from the difficulties posed to us by its *being*. We create *many* dubious things, but no one denies the *intelligence* and *purposefulness* which goes into them.

'As to natural evil, and human suffering, the world's religions have some ideas which may help, even if they cannot answer all our questionings.'

'Such as?'

'Well, one idea is that without opposites or *relativities*—hot and cold, up and down, left and right, pleasant, and less pleasant—there could not *be* manifested existence. There could only be silent, undifferentiated, oneness.

'Another idea is that a mysterious *spiritual falling away, or 'separation,'* took place, aeons ago, on *subtle spiritual* levels. This sad, cosmic, *'fall'* led to this beautiful but challenging world.

'One aspect of this view is that there are *malevolent* forces, sometimes conceptualized as *fallen* angels or demons, which subvert things on subtle levels by creating some of the very difficult conditions, like negativity and disease, with which we have to battle, here, on our dense, PHYSICAL planet.

'These challenging forces, while hard to understand, do, at the same time, play their own parts in creating the great panoply of existence.

'Another idea is that our world is a school for *Souls,* with tough rules.

'Others say that 'free will' means free will, or we would just be puppets.

'This is why *Christ's* teachings were so radical. They *still* are. He taught us that *we* are *not* just instinct-trapped animals, or, mere robotic automatons, forced to return violence for violence every time we are wronged. He said we could use our god-given *free will* in a different way. We could even pray for those who trouble us, rather than hating them, and seeking revenge.

'Others mention that, beyond our sufferings in this world, there's an ultimate *Goodness* to reality, *an amazing, healing Light, and Love.*

'Many of those who have had near death and other kinds of *out of body* experiences report that they experienced amazing s*piritual dimensions* which were *intrinsically* intelligent and *loving*. But, whatever the *deeper* answers to the mysteries of earthly pain and suffering, the claim that *Living*

Nature, and us within it, is just a long series of CHEMICAL COINCIDENCES, is not only demonstrably impossible, it doesn't make sense.

'If we make something clever, Ali, we know that it embodies purpose and intelligence. It embodies *our* purpose and *our* intelligence. But if nature expresses something vastly more ingenious in its dynamic, self-reproducing, self-healing functionality, vastly more intricate than *anything* we can make, we tell ourselves, "Oh, it's just a chemical coincidence?" A few chemicals + a few natural laws + some random swilling around + lots of time are all that's needed? When was this ever a *'scientific'* way of looking at reality?

'No modern thinker, discussing ID, versus Darwin's UD theory, is claiming that some simplistic, Father-Christmas-like god is sitting on a far off cosmic cloud designing Life Forms. No. Just that scientists today are perfectly well equipped to use their *intelligences, combined with modern techniques,* to work out whether they are observing *intelligent and purposeful* phenomena, when they look at certain aspects of nature or RANDOM. But, if we deny all intelligence and purpose in Nature, then we have, perforce, to deny all intelligence and purpose in ourselves.'

'Because we emerge from Nature and must share her qualities?'

'Yes. Then everything we say becomes meaningless, not least the words of those famous science popularizers who, so strangely it seems to me, use their *own, inborn, meaning-making* abilities to *'make meaning' by denying* meaning, their own, amazing, inborn *intelligences* to *deny all intelligence* in the *astonishing!, living!,* cosmic matrix which gives birth to them.

'We use our *intelligence*—not randomly tumbled about parts—-to make Jaguar Cars and Jaguar Fighter Jets. These clever creations of ours, although they are like wooden spoons compared to their *living, feline* inspirations, are designed to evoke, in a small way, the keen intelligence, the predatory purposefulness, and the stunning speed of the big cats of South America. [293]

'Yet our science *'says'* that there is *'no intelligence'* or *'purpose'* to this same, amazing, living nature whose un-admitted designs *we copy?*

'We emerge from her, and we do our own, *purposeful, clever,* best to learn from her, to borrow from her, to copy her designs, but her designs are *not* designs, and, unlike ours, they are *not clever at all?*

'They're just chemical coincidences, which fell-together by—pure—stupid—mindless—probability-defying—entropy-violating—chance, while we, who are, oh-so-clever, can base our own intelligent designs *upon them!*

'We put *millions* of hours into designing *our* robots but mindless chemical coincidences created *us!?* This way of thinking—exemplified by the nonsense put out by *wikipedia* on intelligent design—is not only irrational, it's positively *anti-scientific*. It *damages* science. It leads to a narrow, inquiry-limiting, science-crimping, reality-shrinking worldview, driven by arrogance and ego, not wisdom, let alone common sense.

'Materialism, anti-spiritualism, has such destructive consequences. Not only does it restrict the scope of our inquiries, by insisting that the *spiritual* cannot exist, it cuts us off from the *deeper sources* of things—from the quiet voices of *spiritual* truth within. *Quiet intuitions* which can, potentially, inform our ways with more wisdom, compassion, and commonsense.

'Some say, Alisha, that humanity's perennial debates about origins are not important. They could not be more wrong. Because if there are *no spiritual* worlds beyond this, our visible, physical world, *nothing is sacred.*

'What does it matter, then, if we genetically over-tamper with the species, or poison our skies and seas, or feed ground-up sheep remains to their bovine relatives? It's all just protein, isn't it? There's no *'spiritual'* reason not to do it—because there *is* no soul or spirit.

'If we are nothing more than glorified *'animals,'* whatever that means in the soul-denying materialist scheme, as anima means *Soul,* why not microchip or **ID tattoo** the animal-cattle-people too? You won't need a bank card or passport anymore, just a chip or tattoo. All your personal data visible to THE MINDLESS MACHINE, controlled by real people, behind the scenes.

'If we are just another animal, like the ones we so casually and cruelly chip, brand, and factory farm, if we have no *inalienable spiritual* worth, then we may as well all be branded, and numbered, as Hitler organized in his hellish camps, with the added 'benefit' that a **chip** or **bio-tattoo** can be made to do so much more, even to terminate your physical life, if needed.[294]

'If, as so many now believe, Darwin, and *'the Science,'* have shown us that there is no *God, or Higher Power,* then, with no true meaning to anything, we are, in theory, free do as we wish. No *accounting,* no *karma,* no justice, no *meaning.* The *few,* who may be able, can even strive to be Nietzsche's *Übermensch, or Superman,* for whom *might* alone makes right.

'Within a *spiritualist* ontology, we have the freedom to choose the good, or the ill, but with **consequences,** 'as you sow…' In the materialist scheme, there *are* no consequences. *Our will* is absolute, *no God, no Higher Power, no ultimate Truth, just us, our* ideas, *our* desires, *our* will. *We* are *'god,'* nothing higher than we—and, if we have the power, we can do as we like.

'If we believe that only **MATTER** is real, the repellant trans-humanist idea of turning ourselves into **cyborgs,** part-**robot,** part-human will seem attractive to some. After all, doesn't Darwin teach us that we are just random 'bio-robots'— no more possessed of *'Souls'* than the robots we create?

'The bizarrely unnoticed *difference* being that we put *millions* of hours into creating our robots, but, according to Darwinian UD theory, totally mindless processes made us. *Our minds, Our intelligences* are just cosmic *flukes!*

'In his book *The Science Delusion: Freeing the Spirit of Inquiry,* biologist Rupert Sheldrake explores the delusion that "science already understands the nature of reality, the fundamentals are known and only the details remain to be filled in." But, he argues, our sciences are now,

> 'being constricted by assumptions that have hardened into dogmas. The **'scientific worldview' has become a *belief* system** [rather than an open minded inquiry]. All reality is material or **PHYSICAL** [and all evidence which tells us *different* is mocked, ignored or denied]. The world is a **MACHINE** [which, unlike other machines, 'accidents' created], made up of **DEAD MATTER.**
>
> 'Nature is purposeless [the Lion hunts for no reason, Daffodils grow for no reason, Caterpillars transform into beautiful Butterflies for no reason, Bach composed sublime music for no reason, Picasso painted for no reason, Einstein discovered E=mc2 for no reason]. C O N S C I O U S N E S S is nothing but the physical activity of the [accidental] brain. ['We're all zombies. Nobody is conscious.' Dennett.] Free will is an illusion ['You're nothing but a pack of neurons,' Crick.] God exists only as an idea

in human minds imprisoned within our skulls [and cannot be a **real, direct, personal** *spiritual* experience[295]].

'Sheldrake examines these dogmas, and shows, persuasively, that science would be better off without them: freer, more interesting, and more fun. In *The God Delusion* Richard Dawkins used science to bash God, but here Rupert Sheldrake shows that Dawkins' understanding of what science can do is old-fashioned and [is] itself a delusion.' [296]

'In *The Guardian,* Mary Midgley commented:

'We must somehow find different, more realistic ways of understanding human beings – and indeed other animals – as the **active wholes** that they are, rather than **pretending to see them as meaningless consignments of chemicals**. Rupert Sheldrake, who has long called for this development ... shows how **materialism** has gradually hardened into a kind of [anti-spiritual] anti-Christian principle, claiming authority to dictate theories [like UD] and to veto inquiries on topics that don't suit it, such as unorthodox medicine, [intelligent design, psychic research] let alone religion.

'The 'science delusion' of his title is the current popular confidence in certain **fixed assumptions** – the exaltation of today's science . . . as a final, infallible oracle preaching **a crude kind of MATERIALISM** . . . His insistence on the need to attend to possible wider ways of thinking is surely right.'[297]

'Ah, yes. Far from being scientific, **materialism**—radical-atheism—is a dead end. It restricts our scientific inquiries and it diminishes science by denying or excluding data it cannot account for. Fortunately, more open minded thinkers, like Chris Carter, take the trouble to look at the plentiful modern data on *psychic* phenomena and *Life* after death, and they show, with many examples, that $C\ O\ N\ S\ C\ I\ O\ U\ S\ N\ E\ S\ S,\ and\ M\ I\ N\ D$ are not, ultimately, BRAIN dependent. Their work falsifies the materialist paradigm.

'Today, we've discussed the tremendous *intelligence* in how living things are put together, but, we made no attempt to *define* intelligence.'

'Okay, so how do *you* define it?'

'As an *ontological* quality, as a *basal* quality of *reality, of All that is*. For the spiritually inclined, as an attribute of the *Divine*. It is not physical but we all recognize it when we see it: in the mind of an Einstein, in the design of a Car, in Life's Codes, in the circulation of our blood, in the endless work of our beating hearts, in the dynamic, 24/7 functioning of *all* Living things.

'We cannot *see* nature's intelligence physically, not even in ourselves, because it's not 'PHYSICAL,' but it *pervades* all, and it *informs* all.

'Without it, nothing would work. Think of the bat's amazing sonar or the bombardier beetle's incredible weaponry. Think of birds' feathers and birds' lungs. Think of the magical transformation of the green caterpillar into the multi-hued butterfly, all masterpieces of bio-art and bio-engineering. Think of the fascinating *smart-materials* functionalities of the atomic elements, like carbon, and the complexities of Life's DNA Codes

'You mean Nature is *spiritual intelligence* made *visible?*'

'*Yes* . . . we have no reason to think that the *spiritual* quality, we call *intelligence,* just randomly arises from *'mindless atoms'* bumping and clumping together, before, eventually, becoming human-shaped, and, voila: *'intelligence.'* No. The extremely clever functionalities of those same atoms is an expression of the *same, dynamic, 24/7, intelligence* present in *all* living things, the *same intelligence* from which all things arise.

'We also recognize intelligence in our abilities to respond to reality with evaluation and action, a characteristic we share with animals, even plants.[298]

'Yet, according to science today, this *universal quality* is nothing more than an accidental byproduct of an accidental brain. Really? No. If we consider, we soon realize that *intelligence* is not just a quality of our minds, but a *basic* quality of reality itself—*of All that Is*—as many spiritually realized people, including many NDE'rs, and other out of body experiencers have found. P. M. Atwater describes near death experiencers who,

> 'felt awed by the wonderment of *a blackness [or, at times, light] that appeared intelligent,* emoted feelings and instilled in the experiencer a sense of peace and acceptance.'[299]

'Anne Horne commenting on her NDE, in *The Emotion Code,* says,

'I was going back, home, and on my way, there was a light, a tunnel. I felt like I was being pulled by my heart toward a wonderful place. In that moment, I was encompassed by all this ***innate intelligence*** and ***tremendous love.*** And I just wanted to go home. It was fantastic.' [300]

'What, too, of *spiritual* qualities like *love and truth?* The MINDLESS MATTER of the materialist belief system can know nothing of *love and goodness,* or *beauty and truth.* Yet, *we,* do know those qualities . . .

'But why, Ollie, do these amazing *Soul* qualities and conditions, like *Awareness, like Feelings and Desires, Love and Joy, Will and Intelligence, Curiosity and Creativity, our love of Beauty, Goodness and Truth,* exist at all?'

'No one knows. All we know is that these are *all* qualities of *Life-being-Existence, of* **Reality Itself**—which ***Just Is***. What we *can* say, though, is that what we think of as *intelligent Subjectivity* and *intelligent Agency,* which we, and other beings have, would not work without these qualities.

'If we had *awareness,* for example, but could not respond *intelligently* to reality, we'd be *less* sentient than a basic microbe, which can do both.

'Everything we do, imagine, and create, is a response to, and an expression of our *subjectivity,* and our *experiences,* all of which are informed by our ***awareness***, which is pervaded *by **intelligence**, by curiosity, by* ***creativity***, *by* ***love***, *by* ***desire***—in relation to our ***awareness-held*** experiences of pleasure and pain, hot and cold, beauty and ugliness.

'Does reality—*honestly*—consist in 'DUMB' MATTER alone? Or, does it consist also, as our experience constantly tells us, *in* **inner** *aspects, Mind and* **Soul** *aspects,* **Subjective** *aspects,* **Intelligent** *aspects,* **Spiritual** *aspects?*

'We cannot physically see the *subtle, spiritual realms* of the *panpsychic* vastness in which we live and have our being, nor how *It/S/he (ISH), in all ISH's stupefying power,* gives rise to the *Soul-Life-in-*FORMS of the Plants, Animals and People, the awesome Planets, and magnificent Stars beyond.

'Most of us can only see *ISH's outer* SHAPES. We, moderns, can, sadly, no longer see into the *Soul and Spirit* of things. We can see only the OUTER FORMS of the rocks, waters, trees, plants, animals and people, which clothe

the Souls and Spirits of these things. Yet, with developed *spiritual* eyes, we can start to see so much more.

'This is why, for the world's first peoples, for the yogis and the sages, for the shamans and the mages, the beautiful worlds of nature, and the vasty, starry tapestries beyond, are all parts of the *Great Mystery,* the *Great Spiritual Mystery.* They are its *beautiful, living, manifestations.*

'If we *truly* wish to connect with the *deeper* dimensions of things, we must look *within,* not just outwards with our telescopes, or downwards with our microscopes. As Jesus pointed out, 'the *'Kingdom of Heaven'* is *within.'*

'He didn't say, 'Hey, it's *out there* in the sky, and, one day, when you have really powerful telescopes, you'll be able to see it.'

'No, the *subtle, inner,* realms of reality, the more *subtle, less tangible dimensions and vibrations* of existence, can, during our physical lives, only be accessed by *inner practices* like meditation, visualization, and prayer. We cannot enter them solely with our physical senses or their extensions.

'Materialism, as it sees things one dimensionally, as it denies the *Soulful, Creative, Subjective* dimensions of reality, cannot agree that any such *inner* ways of knowing are possible, nor, indeed, that there is anything more than the mindless coincidences of its deathly, de-souling, system to know.

'For materialism, there are only ever the OUTER and the MECHANICAL to explore. There are just *'accidental zombie'* SENSORS (eyes, ears, taste, touch) which *'purposelessly'* report to the *'accidental brain.'* And the resultant Human Being with *wishes and desires, feelings and dreams,* is not truly a living *Subject, or Soul* at all, just a MEANINGLESS OBJECT.

'A deluded *'it,'* mistakenly believing itself to be anything more than a zombie bio-robot, a pointless "pack of neurons," as Francis Crick had it.

'Never mind that beyond materialism's irrational iron curtain of the mind, where all talk of intelligence and design are thought crimes, no robot *ever* falls together by random—probability-defying—entropy-violating—luck.

'But the dreary, science-restricting, inquiry-limiting thinking which currently controls our sciences, cannot acknowledge the *intrinsic intelligence* in living things. For, to do so, would lead to its collapse as a thought system.

'MATERIALISM can never allow that there is any *true, inner, Being, or inner Soul, or Spirit* to know. Of course not. In its *grim **de-souling*** schemes, there can be no *living Souls* who gaze at us from friend or lover's eyes. There can only be bio-ROBOT-zombies, who may *think* they have free will, and *real* loves and *real* desires but, really, they are deluded.

'Because the 'unintelligently-designed,' Darwinian human of current scientific thinking has no true *Subjectivity,* no true *Soul,* no true *'I-Am-That I-Am' Beingness?'* Remarked Alisha.

'Yes, no mere ROBOT can have a genuine *'Me-You'* relationship, because it is all TAILS, and no *Heads.* All OBJECT, not a true *Subject or Self.*

'Nothing like you and me, all MECHANISM, and *no Soul.* No-one *in* to truly *feel,* no-one in to truly *know,* because in the *de-intelligenced and de-souled* universe of **materialism** no one *is* ever truly at home.

'Yet, no **robot** or AI MACHINE, no matter how seemingly clever, will ever have *true* free will, because it will have no true *subjectivity or Soul.* It may be made to *look* as though it is aware—but it won't *truly* be. It will simply be a *mindless* automaton which we, idol makers, and idol worshippers, so in love with our own cleverness, are all too ready to worship, forgetting that it's just an algorithmic idol, we ourselves made.

'Sure, a ROBOT may be designed to *look* as though it cares, as though it wants to love, or laugh, or to play, but it will be a hollow mockery of anything truly *Living or Soulful* because it will have no, true, *'I-Am-that-I-Am-Being.'*

'Behind it will always be the *human* intelligence which created it, an intelligence, which though very clever, cannot give it true *Life* or true *free will.* Because, we, *unlike Spirit,* cannot confer *Living Soul* upon it, but only INANIMATE SENSORS and lifeless CPUs.

'No ROBOT can ever be truly *sentient*.

'A MICROPHONE is not a living *Experiencer* who hears.'

'An ELECTRIC SPEAKER is not a living *Subject* who speaks.'

'A LIGHT SENSOR is not a *Soul* who truly perceives.'

'A ROBOT'S CPU can only recycle the programs coded into it—*by* **us***.*

'No ROBOT can truly choose. It can only *react,* using the rules and the ideas, the codes and the algorithms programed into it—*by* **us***.*

'But surely, Ollie, everyone understands the differences between AI Robots and *living Beings—sentient—Soulful—Living Beings.'* Smiled Ali.

'Ah, beloved friend, if only it were so. Too many do not seem to understand the difference. This is why some people even talk of giving **robots** 'human rights'— maybe because they think *we* are just robots.

'This is why there is talk of 'artificial intelligence' taking over the world—forgetting that AI is *our* creation, and, all we have to do, is pull the plug, unless we foolishly allow 'the machine' to be able to stop us from doing so.

'That would be *our* fault, not the "AI's." Yet if too many people, truly come to, believe that the sacred conditions which we understand as *Sentiency and Soul, Mind and Consciousness,* are mere accidents of mindless matter, then they *may* start to argue, as some already do, that we *are* no more fundamentally *'sentient' or 'alive'* than **robots**.

'Shockingly, according to mainstream, western science today, we do not really have real *minds* or real *souls*. No, we just have pointless, chemical-accident brains, which work, for *no point at all!* This grim orthodoxy which makes science look *irrational*. Is this why Stephen Hawking argued that philosophy is dead? Because science had killed it? If this were true, Ali, it would be like arguing that science has killed reason and common sense.

'Philosophers, he said, 'have not kept up with modern developments in science and... philosophical problems can [now] be answered by science...'[301] Really? No, brilliant as he was, he was very wrong on this.

'Philo.sophia means the *love of wisdom*. But our science, clever as it is, is mostly ***technical*** knowledge. It is not philosophy—nor is it meant to be. It cannot, always, even tell us what is *true,* let alone what is wise and good.

'Those famous modern intellectuals who preach that Darwin's anti-spiritual theory of Life is *'right,'* despite it so clearly not being right, use their *Soulful caring, their love and their desire,* all non-material, Soul qualities, for what is *good, beautiful, and true,* also *spiritual* qualities, to argue for total soullessness and total cosmic unintelligence.'

'That seems so self-contradictory to me...' Mused Alisha, looking sad.

'Of course it is. For, if there were no *intelligence* in Nature, how could we, who emerge from her, *be intelligent,* and how could she be *intelligible?*

'Science is, above all, mean't to be a *rational, logical* activity. How, then, can we advise thoughtful young people to go anywhere *near* certain areas of this once logical discipline which tells them, in the name of reason, that they are 'nothing but packs of neurons,' that there is no intelligence or *teleology* (purpose) in nature, and all living things, including we, are mere coincidences of chemistry. Am I making this up? Look up 'intelligent design' in wikipedia. Unintelligent design, or UD—although they don't call it that—[as that would make the nonsense too obvious?]—is the mainstream, anti-supernaturalist, anti-spiritual idea that, 'undirected,' unintelligent, chance-based processes can explain all.

'Yet, everywhere we *look,* we see *tremendous* intelligence and *incredibly* purposeful functioning in nature. There's not one atom that's not *doing* something—that's not *functioning*—even if all it is doing is helping to make a rock. Yet, without that clever functioning, of *all* of nature's components, including atoms, there would be no atom, no rock, nor any *thing* at all.

'*Teleology,* or purposeful functioning, is the very *warp* of reality.

'There can be no movement, not even of an atom, without *force*. Because without force nothing would ever move. There'd be *eternal stillness.*

'But, as soon as there is *movement* of any kind, even of a quark, there is *functioning or work.* This implies *purpose,* because *things always work or function for a reason.* Reason always points to *motive or motivation,* which always points to a *Being or Beings* with reasons and motives, which always implies *Subjectivity* and *desire,* even to the level of the *Spiritual and the Divine*—which implies *love*—because only that which is loved is desired.

'But without an amazing *intelligence and wisdom* to all of nature's movements and laws, at all levels and dimensions, nothing would ever work. Which points not only to **Will**, and to **Love**, but to great **Intelligence** at work in nature. This is why holistic reality is indivisibly made up of

⊙ *Intelligence–Creativity* ⊙ *Love–Desire* ⊙ *Will–Action* ⊙

'It is from this mysteriously inter-weaving *trinity of essential qualities,* imaged, micro-cosmically, in us: **Head** (Mind-Intelligence-Creativity), **Heart** (Joy-Courage-Love-Compassion-Desire-Curiosity), **Belly** and limbs (Will-Power-Purpose-Action), that holistic reality is created. This is why, if the cosmic matrix of our birthing had no ***intelligence***, neither would *we*. If it had no *love or desire*, neither would *we*. If it had no **power or purpose**, neither would *we*. This is why *panentheism* makes sense.

'Panentheism?'

'Panentheism is our intuitive knowing that a *mysterious, all-pervading, Cosmic, Animating, Force, which* **'*Just Is*,'** not only gives rise to, and *pervades* Nature, but, also, timelessly extends beyond it.'

'You mean god *immanent* and god *transcendent* in religious language?'

'Yes. The mysterious *Spirit (It/Is/I/S/he/You/Are, 'ISHYA')* which *gives rise to all, IS within All, and beyond All*. Panentheism reflects the insight that *we, humans,* are microcosms of the macrocosm. Intriguing humans, micro-gods, hologramatic reflections of the whole, imbued with *intelligence, love, desire, curiosity, creativity and will*—made, in a sense, in *ISHYA's* image.

'Our challenge, though, is that, unlike the ʀᴏʙᴏᴛs we make, we *do* have free will. We, are, *genuinely,* free to choose the good, the beautiful and the true, or their opposites. We turn towards *God, Essence, Presence, the Now, the Tao, Buddha Nature, Supremely Intelligent, Creative Beingness, Brahman, Allah, Jehovah, Wakan Tanka, Ish, Ishwya, AUM, the Great Spirit or, simply, the Great Mystery, the labels don't matter*, not merely because our ancestors were afraid of thunder and lightning, and could not explain them, as some claim, but because something in us hears a *deeper* call.

'Something inside which remembers that we are born of this same *Spiritual Nature* to which humanity gives its traditional religious names. We are born of *ISH's Shimmering Living Fields*. We come from *AUM or ISH* and we return to *AUM*. This is why we *love Truth* and why *we* seek *Truth*.

'This is why our quests for scientific and philosophical truths are part of our *very nature*. This is why our attempts to discern the good, the beautiful and the true, and to live by our discoveries, are parts of our *True Nature*.

'It is this same, *subtle,* **Living-Loving-Light–Intelligence** of our deeper **Being** which, quietly, curiously, creatively, looking through our eyes, pursues science, philosophy, religion, or art, seeking, ever-onwards to know, and to go, ever deeper into the *mysteries* and the *truths* of nature, in all her aspects, ᴏʙᴠɪᴏᴜs and *subtle,* superficial and *deep*—and, for those of us so drawn, for spiritual fulfillment, self-realization, mystical, and divine union.

'Some, *subtle, inner force,* which quietly whispers to us that we are *more* than mere cosmic accidents, that if that's all we were, we wouldn't care. We came into this world, as Wordsworth beautifully put it, *"trailing clouds of glory."* Something in us remembers, and calls us home. Our lives are not merely "tales told by idiots, full of sound and fury, yet signifying nothing," as Shakespeare's tragic Macbeth cried out, but meaningful.

'As Anita Moorjani discovered in her amazing NDE, we are *spiritual* Beings having temporary physical experiences. We are, as she found, to her astonishment and her delight, *magnificent* beyond our own imagining, not in some childish or egotistical way, but in our very *essence,* as *spiritual beings.*

'It is this same, *quiet, diamond-spark, essence-presence-awareness in us, subtle-softly-ever-aware of its ultimate Source,* which, when we are willing to attune to its quiet voice, its whispered intuitions, gives us our *own, deepest values of the—most good—the most precious—the most beautiful—and—the most true.'*

'Ah, these are all such deep mysteries...' Alisha sighed.

'Yes, they are. Perhaps we can go into them some more one day. But, for now, as our quiet walk by this gentle river, in this beautiful valley, comes to a close, let me thank you, beloved friend, for giving me a chance to share these precious ideas with you.'

'You're very welcome, it has been very interesting.'

'I just wish, my dearest Ali, that, gradually, more and more people will come to see that materialism, or radical atheism, and all purely materialistic accounts of Life, like Darwin's, while currently fashionable, both as science and philosophy, don't work.

'Not only this, but they instill nihilism and despair wherever they take root. Within science, *especially* within science, they block our deeper progress, because they block a deeper understanding of *who and what* we truly are, and a more *open-minded,* and more *open-hearted* kind of inquiry.'

Valley quietening, skies darkening, stars sparkling, we headed home.

ALPHA AMEN AUM AMIN OMEGA

Epilogue

That's it, at last, it's done. I can go away. ... "Hold on, just a few more things to say." ... No... Enough. Fifteen long years on all this...'

Ah, beloved Reader, and you really are, if you have stayed with me this far, I wonder what you think? Are you persuaded? Perhaps you were always *spiritually* inclined, but not so sure about Darwin? Or, were you a fan of Darwin and his ideas, now not so sure?

So, IS EVOLUTION TRUE?

Well, who am I to say that the *Great Spirit* [or your preferred words] only *creates* things, *Atoms and Galaxies, Cells and Shells, Plants and People,* but never *evolves* them? Perhaps our amazing universe is a vast thought experiment, in which creation and evolution go hand in hand?

I do, though, question Darwin's *"No Spirit, No Intelligence needed,"* version of evolution. Yet, if his idea of a universally shared descent—microbes to elephants, particles to people—had been *supported* by the fossils, as he hoped, with huge numbers of transitional forms, along the way, why would I argue? The data would speak for itself. The trouble is, it is quite clear by now that, evolution, as *he* described it, did not happen: randomly, as he said, or intelligently, as others claim. So, where are we?

Firstly, the scriptures are not science, they serve spiritual truths.

Secondly, there is no need to believe in Darwin anymore. We did not need him to tell us that Life on Earth has changed vastly, over vast time. The Cambrian Big Bang, the amazing Dinosaurs, and so on, all happened, and we knew these *before* Darwin. But they never matched his predictions.

So what have we lost? Not a lot, apart from Darwin's now, tired old *"Just So"* story, which, while *scientific in intent,* turned out not to be correct. Trying to force Life's fascinating mysteries on to the unviable limb of materialism, and its unhappy partner, Darwinism, has not worked.

"Dear Darwin, you hoped to solve Life's mysteries. How did it all start, and unfold? You answered: "First, microbes arose, by chance. Then, a gradual descent—with such vast changes—as to morph into all Life." Trillions of rAn*d*om changes, which your modern followers believe were caused by trillions of (*rare!*) ra*n*dOm, genetic typos—with no realistic explanations for the origins of genes, nor for any organism, in the first place.

"On the Origin of Species, you said, was just an *"introduction"* to your ideas, which you would later fill out. Meantime, you hoped that *proof* for your hunch, for your *bluff,* would be found in a vast number of transitional forms—fossil and living—except, no such proofs were ever found.

"You hoped you had cracked Life's codes. How glorious that would be. But what about the facts? Sadly, it seems you could not allow the facts to spoil your brilliant theory: Microbes to Worms, Spiders, and Snails. Rats to Bats. Bears to whales. So, you ignored so many of the actual facts."

Ah, humanity! We threw out one childish set of beliefs, the Father Christmas god, the frying pan, and replaced them with an even more absurd set of beliefs. *Life and Mind* were just accidents! The fire! The frying pan, at low heat, may not kill us. The fire will. And, before it does, it will drive us mad. Such a crazy fire of ego, hubris, unreason and illogic—until truths become lies:

> "Nature is *not* intelligent and purposeful, nor does she have intelligent sources." And lies become truth:
>
> "Oh, didn't you know? Dumb chance and "undirected processes" are incredibly creative. They make everything. Look up intelligent design in *wikipedia.* It says so…
>
> "The science is on our side—and the 'science says' that only PHYSICAL STUFF + BLIND CHANCE is real. No *spiritual* causes, no *spiritual* dimensions. Forget about your immortal *Soul*—and Life *after* death. You don't have one—and there isn't any. Get over it."

Ah, yes: "The Science is on our side." It is strange, though, that if we ask our materialist friends to explain to us how Life *could* (ever) arise by mere chance, and what their *scientific* **evidence** for this is?, they will either get blustery, or defensive, or go all vague on us.

If we ask them how a DNA Code—"like a computer program, but far, far more advanced"—could arise by chance, we will be met with hand waving or vagueness. If we ask for the data showing how spiders and their webs, or bats and their sonar 'evolved,' or how caterpillars evolved to transform into butterflies, or how the bombardier beetle's extraordinary weaponry 'evolved'—all by *pure chance?* We'll hear more *"Just So"* stories.

If we share, book, after book, with powerful modern evidences for the *super* natural, and the *spiritual,* including for verifiable communications, from those passed on, they will almost certainly not bother reading them, while claiming that "the science," that Darwin's theory, is on their side: "Life is just an accident which evolved by accident, and Darwin proved it!"

No. None of this is 'science.' It's MATERIALISM. The illogical, science-distorting superstition that mere chance can create all: "Come on, that's impossible." . . . "Ah, but "the science" is on our side." No, it's not true.

So, dear Reader, if you are spiritually inclined, please take comfort, "the science" is **not** on the side of the 'skeptics' of the *spiritual* and the *super* natural. It is *not* scientific to ignore that Darwin's theory never stacked up.

No, ironically, the *real* science is on your side.

The *real* science tells us that Life could *never* start by chance—it is chemically impossible. The *real* science tells us that cells work, very cleverly, to *prevent* random genetic mutations—supposedly, the very engine of Darwinian evolution—with massive changes—microbe to snail, microbe to whale. The real science tells us that there are NO TRANSITIONAL SPECIES, FOSSIL OR LIVING, AT ALL, let alone vast numbers of them.

Like cartoon characters running off a cliff... but not noticing...Darwin's fans keep running... no longer any scientific ground beneath them.

Actually, there never was. They just persuaded themselves, and too many innocent bystanders, that there was. I have been wondering, though, how could all this happen? How could such a **flawed**, and **repeatedly falsified**, set of ideas be treated as 'good science' for so long. Then it dawned on me.

> Like a tragic fairy tale: "Charles Darkwing, amateur scientist, dreaming of scientific glory, had a new theory of Life's history—because the old *religious* ideas were nonsense. So, the people

helped him build his amazing time-machine, all according to his theory, and it crashed, horribly, killing the pilot, horribly.

"He, though, and his supporters, were undaunted. He said, and they agreed: "My theory is *far too good* not to be true. It **must** be true. It *feels* true. I will amend my plans a little, and let's have another go." So they did and another precious life was lost."

Do we really think that if Darwin's ideas had involved anything **remotely practical**, like a new airplane or a new drug—rather than being, primarily, of philosophical and (anti) spiritual interest—we would **not** have dropped them long ago? His theory never flew… it crashed immediately. But, who cared?

It was a great story—so why let the **real** scientific facts, of the fossils, embryology, and so on, spoil that? No one was going to get hurt… It was just a *virtual* time machine. So no one *could* get hurt. We couldn't *literally* go back to the Cambrian seas, or the Mesozoic and the amazing Dinosaurs.

Except, it's not true that no one got hurt. Because of the relentless promotion, as 'science,' of Darwin's failed ideas, despite all the contrary data, ideas which were mostly just philosophy, plus speculation upon speculation—never proven science—millions of people lost their sense of the *deeper Spirit of things,* and their *inner* access to the *Great Comforter*.

Ideas have consequences.

Darwin's *bluff* had massive consequences, also, in the non-science realms of politics, philosophy, psychology, ideology. The Nazis liked Darwin, survival of the strongest! The Marxists liked Darwin—we could create our own, god-not-needed paradises, right here, on earth. Totally materialistic paradises—millions might die in the process, but who cared?

Who cared if what he said was actually true. He'd got rid of God and Spirit and Soul, and that was a good thing, wasn't it? Well, not if *God* and *Spirit* and *Soul* are real, and are the *true, Living Sources* of everything.

Changes over time? Yes. But what Darwin meant by evolution? No. Darwin wanted to be a rock-star scientist. He sort of got his wish. But the rocks never granted him his *real* wish—to be right. No, they proved him wrong—again and again. Still, he didn't seem to care. His theory was, he felt, like Darwkwing's fatal time machine, *'too good'* to be wrong.

This is how his followers have continued. It doesn't seem to matter what the data says, they "know" Darwin was right, and those of his fans who are materialists, "know" that *the spiritual and the divine* cannot be real.

"Let's stick with Darwin, no matter how much nonsense we have to believe, or data to ignore, to continue to believe what we believe." Ah, yes, we can all believe whatever we wish.

> "If we don't agree that Life could **never** start by chance, or that no **CODE**, including Life's Codes, could **ever** arise by mere luck, we don't have to agree—do we? If we say that bears swam and slowly became whales, or rats slowly became bats, or that spiders, and their sticky webs, arose by pure, sticky chance, who are you to question? May we remind you, too, that religion is the 'root of all evil.' This alone will prevent us from taking anything that points to the *spiritual* or to the *super* natural seriously."

If we reply, "Hold on, Hitler, and others, who were not religious, were responsible for just as many, or more, deaths, than many religious wars," our materialist friends will probably either ignore us, or, perhaps, they'll say,

> "Look, how can you believe that some absurd 'bearded-god' makes everything. It's ridiculous. No, we need proper **SCIENTIFIC** explanations for things, not look to outdated religious nonsense."

"Ok," we reply, *"ideology,* of any sort, religious, communist, fascist, any *'ism,'* if taken to extremes, can become destructive, especially if it is used to dehumanize those who disagree. We don't believe in your straw man bearded-god, either. Nor are we promoting any religion, just the amazing, all-pervading, *Subtle,* **Force**(s)*, which, this *second, and every second,* gives (give) *rise to,* and *is, (are),* the *Living* **Source**(s) *of All,* and *the very fabric* of *All that Is, all* `matter,` `all` `energy,` `all` `space,` `all` `time.`

"We live, despite its challenges, in an *intelligent* universe. *Intelligent Mind* is as much a property of reality, as S..O..L..I..D M..A..T..T..E..R, and there are *levels of Mind, Spiritual or Divine Mind,* which make *our* minds seem utterly puny—look at cells, look at any living thing!"

Emotional reasons *not* to believe in a meaningful, *'something more,'* *beyond* this beautiful but often difficult world are many. So much individual and shared suffering, human and animal. It is hard to make sense of all of that.

So, perhaps, at times, it is easier, and less painful, *not* to believe. To be a *'tough-minded 'realist,"* to give up on the *Unseen, on the Friend, and the Comforter.*

But the ***intellectual and scientific*** reasons to deny the amazing, *second, by second, by second, intelligence* which sustains all Living Beings are few, and ever decreasing, *zero* really, as we have seen, again and again.

Wonderful reader, thank you so much for sharing this journey. I wish you so very well. May this work now continue, on the wings of truth and destiny, into our world, with all its follies and its inspirations, where, may it add to the light of wisdom.   A...........................U................................oM

Further Reading

Below are some of the works drawn on in exploring these themes.

On the Origin of Species by Charles Darwin, 1859. Darwin's famous atheistic creation story. Darwin dismissed the **lack of fossil evidence** for his scheme by appealing to poor fossil preservation. But he conceded that those who did not find this idea convincing, **"will rightly reject my whole theory."**

Evolution: A Theory in Crisis by Michael Denton, 1986–2016. Updated. **Must reading** for a more detailed understanding of why Darwin's famous theory has failed to deliver. One of the inspirers of the modern ID movement.

Darwin's Enigma, Luther Sunderland, 1988. Darwin conceded that a lack of transitional fossils showing earlier forms morphing into different ones **would *falsify* his theory**. Sunderland interviews leading fossil authorities who confirm that—in conflict with Darwin's ideas—there is **no fossil evidence** for evolution as he described it. Darwin's theory *was falsified decades ago* but, strangely, most schools and universities take no notice.

Shattering the Myths of Darwinism: A rational criticism of evolution theory, Richard Milton, 1992–2017. Darwinism the *scientific* theory has *failed*. It has been replaced by **Darwinism the *faith***. The *faith* is held to be "fact" while it "totters atop a shambles of outdated... evidence" which should have been questioned long ago. Richard Dawkins and *Nature* magazine took Milton's critiques seriously enough to launch scathing attacks on his book.

Darwin on Trial, Phillip Johnson, 1993. **Bestseller**. Also, *Reason in the Balance: The Case Against Naturalism [Materialism] in Science.*
Darwin's Black Box, The Edge of Evolution, Darwin Devolves, Michael Behe, 1998-2019. Explorations of the intelligent design (ID) hypothesis of *irreducible complexity,* and the extremely limited creative powers of Darwinian evolution. Behe's **unassailable** arguments sink Darwin's ship.

Icons of Evolution: Science or Myth? Jonathon Wells, 2002. Examines the surprisingly few alleged 'proofs' for macro-evolution. He shows that they are all misleading, misrepresentations of evidence, or downright fraudulent.

The Naked Emperor: Darwinism Exposed, Anthony Latham, 2005. Discusses the shocking lack of real world evidence for (large scale, long term) evolution as Darwin imagined it. Excellent, readable, concise book.

Billions of Missing Links: A Rational Look at the Mysteries Evolution Can't Explain, Geoffrey Simmons, 2007. Examines some of the many biological functions "that came about **all at once, entire**…with **no preceding links**, no subsequent links, **no 'sideways' links**. ... **Nature contains only LEAPs, not links.**" Only *super natural intelligence* can truly explain Life's complexities.

Signature in the Cell: DNA and the Evidence for Intelligent Design, Stephen Meyer, 2009. Codes, including **DNA CODES** *cannot* write themselves. It should be obvious. Best seller, **powerful**—logically irrefutable.

Intelligent Design Uncensored, William Dembski and Jonathan Witt, 2010. Using *modern, scientific* techniques, shows how to determine whether the *apparent design* in nature—which we can all see—is random or *meaningful.*

Alfred Russell Wallace: A Rediscovered Life, Michael Flannery, 2011. At first, evolution was referred as 'the Darwin and Wallace theory.' But Wallace, in time, unlike Darwin, saw through materialism and endorsed *spiritualism.*

Darwin's House of Cards: A Journalist's Odyssey Through the Darwin Debates, Tom Bethell, 2016. Having met many influential figures in the evolution world, he concludes that, although Darwin's theory is still touted, it is now *"**disintegrating** under an onslaught of new scientific discoveries."*

Evolution's Final Days, John Morrison, 2019. Excellent, brief book.

Taking Leave of Darwin, Neil Thomas, 2021. Once a convinced Darwinist, he explains why he now considers Darwin's macro theory to be falsified.

Darwin's Bluff: The Mystery of the Book Darwin Never Finished, Robert Shedinger, 2024. Darwin promised he would make good his bluffs in *"On the Origin of Species,"* on the powers of natural selection to evolve all Life, in a later book. Darwin gave up on his *later* book when he realized he could not back up his ideas. He could not find the needed evidence. Darwin's draft MS *was* published in the 20th century but added little to *On the Origin of Species.*

Journal of the Discovery Institute, **Evolution News**. Regularly exposes how the public continue to be misled into thinking Darwin's ship still floats.

Science and the Near Death Experience, Chris Carter, 2010. How near-death experiences or NDEs provide glimpses of an awaiting *afterlife. Science and the Afterlife Experience: Evidence for the Immortality of Consciousness*, Carter, 2012. *Mind* and *Soul* exist beyond biology. *Science and Psychic Phenomena: The Fall of the House of Skeptics,* Carter, 2012. Evidence that 'telepathy, clairvoyance, precognition, and psychokinesis are all real.'

277

Author's brief Answers to Evolution Questionnaire.

1. Until now, have you believed Darwin's theory to be true? 'Once, yes. But, when I began to look deeper, I discovered, to my surprise, that, despite all the claims, the evidence for Darwin's ideas did not stack up—at all.'
2. If Darwin's theory is *not* true, will it shock you, or please you? 'Neither. I'm interested in what's *true*. I'd be ok with Darwin's theory being true, if it *was* true. I would question, though, that such a complex process could be **unintelligent.**'
3. Yes, I think *Life and Mind either* have intelligent causes, which, mysteriously, impossibly, *'just are,'* or they arise by *mindless* causes which mysteriously, impossibly *'just are.'* 4. *'Impossibly'* because how come there is anything-at-all, including a *'Cause of Things,'* rather than nothing-at-all, and *'No cause of things.'*? Do I agree? Yes. Existence has a Cause (or Causes) which *'just Is,'* (or *'just are.'*) 5. We cannot say **why** this is so, but we *can* try to work out if it, or they, are more likely to be mindless or intelligent. **Intelligent** makes more sense to me.
6. Do you think **materialism** is true and *mere chance* is highly creative? 'No. I'm not a materialist, nor do I think that **mere chance** is highly creative.'
7. Has science proved that belief in the *spiritual* is just wishful thinking? 'No.' Or, is science simply a range of methods for investigating reality—be it solely *physical,* as materialists say, or also *spiritual,* as spiritualists say? 'Yes.'
8. Do you think scientists are clever enough to work out that *Existence* is **mindless and unintelligent,** at its core, and in its **causation,** but **not** clever enough to work out that this may not be true? That, actually, **Existence-Being-Life** is the **outward** expression of an *inexpressibly powerful, inner,* **Spiritual Intelligence?** 'Yes, I think scientists are clever enough to work this out.'
9. Is science just a matter of majority views holding sway, 'No.' or do minority views, and alternative hypotheses, also have a role to play? 'Yes.'
10. Is it ok to let go of a theory, like Darwin's, even if there is nothing better to take its place? 'Yes.' Or, is it better to stick with the old theory, than to follow newer evidence, and newer thinking, which falsifies the old theory? 'No. There's never a good enough *scientific* reason to ignore data which *falsifies* an old theory. When we ignore the data, when we ignore the actual, real world evidence, it is usually for political, emotional, or philosophical reasons—not ones of science.'
11. Can we take anything of value from an unknown author? 'Yes.' Or should we always follow the authorities and the majority views of the day? 'No. I'm more interested in an author's ideas than in their paper qualifications—especially in a field which is as much about ideas and philosophy, as it is about science. In this case, I have relied on others, far more qualified than I, for the science discussed.'

End Notes

[1] Dawkins, R. *River Out of Eden: A Darwinian View of Life*.

[2] Rob Merkx, teacher in the Diamond Approach, aka Ridhwan, of A.H. Almaas. Rob: "Presence lives in the now… Life…[is] an ongoing miraculous unfolding. It doesn't mean I always feel good. It simply means there is more trust that all the experiences that life brings are precious." His wish is "that more and more people discover presence" and "start living from a bigger perspective that is not guided by fear, survival instincts or the misunderstanding that we are separate from each other or nature."

[3] Rudyard Kipling's **Just So Stories**, for children, 1902, humorously "explained" various animal features, such as the elephant's trunk or the leopard's spots. Critics of Darwin have pointed out that, while attempting to be scientific, many of Darwin's ideas, like bears-to-whales, are no better than Kipling's **Just So Stories**, with the key difference that Kipling was being playful, while Darwin really wanted us to believe in his claims. It was all very well for Darwin to say, "Can we not imagine this or that incredibly complex biological mechanism evolving in tiny little increments?" But Darwin forgot that imagining things, even if seemingly plausible, does not make them true. Science is not just based on imagination, useful as it can be, but also on evidence, experiment, and common sense.

[4] Lennox, J. 2009.God's Undertaker*: Has Science Buried God?*. Lion. P. 70.

[5] Ross, H. 1995. *The Creator and the Cosmos*. Navpress. p. 117.

[6] Penrose, R. 1989. *The Emperor's New Mind. Oxford University Press.* p. 344

[7] Penrose, R. 1988. *The Cosmic Blueprint. Simon & Schuster,* p. 203.

[8] Das Wesen der Materie [The Nature of Matter: "There is no matter as such. All matter originates and exists only by virtue of *a force*. . .mind is the matrix of all matter."], speech at Florence, Italy (1944) (from Archiv zur Geschichte der Max-Planck-Gesellschaft, Abt. Va, Rep. 11 Planck, Nr. 1797)

[9] Goswami, A. 1993. *The Self-Aware Universe: How Consciousness Creates the Material World.* Jeremy P Tarcher. Capra, F. 1975.The *Tao of Physics.* Shambhala. Davidson, J. 1992. *Natural Creation or Natural Selection? A Complete New Theory of Evolution.* Element. See also Blavatsky and Steiner.

[10] Dawkins, R. *River Out of Eden: A Darwinian View of Life.*

[11] Lesiola, M. 2018. *Heretic: One Scientist's Journey from Darwin to Design,* Discovery Institute Press.

[12] www.catholic.com/magazine/online-edition/does-it-matter-that-many-scientists-are-atheists#

[13] Darwin to J.D. Hooker. 2.2.1871. Emphasis and text in square brackets added.

[14] Davies, P. 1998. *The Origin of Life.* Penguin.

[15] Orgel, L. 1973.*The Origins of Life. John Wiley;* p. 189. Emphasis added. Learnt of Orgel's work in S. Meyer's *Signature in the Cell.*

[16] Denton, M. 1985. *Evolution: A Theory in Crisis.* Burnett Books, p. 260, emphasis added.

[17] Shroeder, G. L. 1996. *Genesis and the Big Bang.* Bantam Doubleday Dell

[18] *Wikipedia,* at time of writing.

[19] Dembski, W and J. Witt. 2010. *Intelligent Design Uncensored. IVP, p. 66.*

[20] Ibid. p.67, text in square brackets and emphasis added.

[21] Dembski, W. 2006. *The Design Inference: Eliminating Chance through Small Probabilities. Cambridge Studies in Probability, Induction and Decision Theory.* Cambridge University Press.

[22] Ibid., p. 71.

23 Ibid. p. 71. Text in square brackets added.

24 Denton, M. 1985. *Evolution a Theory in Crisis.* Burnet Books, p.250, emphasis added.

25 *Ibid. p 328.*

26 Morrison, J. 2019. *Evolution's Final Days: The Mounting Evidence Disproving The Theory of Evolution (problems, myth, hoax, fraud, flaws)* Emphasis added.

27 Sunderland, L. 1988. *Darwin's Enigma: Ebbing the Tide of Naturalism.* Emphasis added.

28 Gould, S. J. 1985. *"Not Necessarily a Wing" Natural History, October, pp. 12, 13.* Text in square brackets added.

29 Lovtrup, S. 1987. *Darwinism: The Refutation of a Myth.* Croom Helm Ltd., Beckingham, Kent, p. 275, text in square brackets added.

30 Michael Flannery. 'Was Darwin a Scholar or a Pitchman?' *Evolution News,* October 20, 2015.

31 C. Darwin, *On The Origin Of Species*, Emphasis, text in square brackets, and some extra paragraph spacings added.

32 Davidson, J. 1992. *Natural Creation or Natural Selection – A Complete New Theory of Evolution.* Element; p. 13.

33 Denton, M. 2002. *Nature's Destiny: How the Laws of Biology Reveal Purpose in the Universe.* Free Press. p. 342

34 C. Darwin, *The Origin Of Species*, Chapter X, "On the Imperfection of the Geological Record. Emphasis added.

35 Bethell, T. Sept. 18, 2013, Discovery Institute, and in *American Spectator*

36 Bethell, T. 2016. *Darwin's House of Cards.* Discovery Institute, emphasis and text in square brackets added.

37 Demolishing Darwin's Tree: Eric Bapteste and the Network of Life. *Evolution News.* 9.9.13. Emphasis and text in square brackets added.

38 Johnson, P. 2010. *Darwin on Trial.* IVP Books, p. 17.

39 Gould, S. J. 1980. *The Panda's Thumb.* p. 181-182)

40 Simpson, G.G. 1944. *Tempo and Mode in Evolution.* Columbia University Press; pp. 105, 107. Emphasis and text in square brackets added.

41 Darwin, C. 1859. *On the Origin of Species by Means of Natural Selection or the Preservation of Favoured Races in the Struggle for Life,* reprint of 6th edition (John Murray, 1902), p. 341–342. Emphasis added.

42 Goldschmidt, R.B. 1940. *The Material Basis of Evolution.* Yale University Press; p. 390.

43 Sunderland, L. 1988. *Darwin's Enigma: Ebbing the Tide of Naturalism.* Emphasis added.

44 Ibid

45 Davidson, J. 1992. *Natural Creation or Natural Selection? – A Complete New theory of Evolution.* Element

46 Latham, A. 2005. *The Naked Emperor: Darwinism Exposed.* London: Janus, p. 39.

47 Gould, S.J. *Wonderful Life. 1990. The Burgess Shale and the Nature of History,* New York: Vintage, pp. 59–66; emphasis and text in square brackets added.

48 *Ibid.*

49 Denton, M. 1985. *Evolution: A Theory in Crisis*. Burnett Books.

50 Meyer, S. C. 2013. *Darwin's Doubt: The Explosive Origin of Animal Life and the Case for Intelligent Design.* Harper One. Book description.

51 Darwin, C. 1996. *On the Origin of Species*. Oxford University Press, p. 249.

52 *Darwin's Doubt,* amazon review, C. Luskin. Emphasis added.

53 *Darwin's Doubt,* amazon review, C. Luskin. Emphasis added.

54 Battson, A. *On the Origin of Stasis by Means of Natural Processes.* Battson's website. Emphasis and text in square brackets added.

55 Lennox, J. 2009. *God's Undertaker.* Lion; p. 143, emphasis added.

56 *Ibid.*

57 Milton, R. 1992. *Shattering the Myths of Darwinism,* Inner Traditions, emphasis, text in square brackets added.

58 Battson's interesting suggestion regarding what we should be looking at in natural history.

59 Latham, A. *The Naked Emperor: Darwinism Exposed.* London: Janus, p. 61 emphasis and words in square brackets added.

60 Battson, A. 1997. *Facts, Fossils, and Philosophy.* Author's website. Emphasis and text in square brackets added.

61 Darwin, C. 1859. *On The Origin of Species. Ch.VI.*

62 Darwin, C. 1859. *On the Origin of Species by Means of Natural Selection or the Preservation of Favoured Races in the Struggle for Life.* Emphasis and capitalizations added.

63 Latham, A. *The Naked Emperor: Darwinism Exposed.* London: Janus, p. 76, emphasis and words in square brackets added.

64 Ibid, p. 77 emphasis and words in square brackets added.

65 C. Darwin, *On The Origin Of Species*, Emphasis, capitals, bold etc added.

66 *https://evolutionnews.org/2018/08/inexplicable-species-and-the-theory-of-evolution/*

67 *https://evolutionnews.org/2018/08/inexplicable-species-and-the-theory-of-evolution/*

68 Article by J. Wells. *https://evolutionnews.org/2008/04/is_the_science_of_richard_dawk/*

69 Letter to J.F.W. Herschel, 23.5.1861. www.darwinproject.co.uk

70 Johnson, P. 2010. *Darwin on Trial.* IVP Books, p. 16, emphasis and text in square brackets added

71 Dembski, W and J. Witt. 2010. *Intelligent Design Uncensored.* IVP, p. 72.

72 Monod, J. 1972. *Chance and Necessity.* Collins. pp 134–135, text in square brackets added.

[73] Latham, A. *The Naked Emperor: Darwinism Exposed.* London: Janus, p. 13. Emphasis added.

[74] Hoyle, F and N.C. 1981. *Wickramasinghe, Evolution from Space. J.M. Dent & Sons.*

[75] Ibid. pp. 141, 144, 130

[76] Eastman, M & C. Missler. 1995. *The Creator Beyond Time and Space. Emphasis added.*

[77] Gitt, W. 2006. *In the Beginning Was Information: A Scientist Explains the Incredible Design in Nature.* Master Books, p. 124; emphasis and text in square brackets added.

[78] Lennox, J. *2009. God's Undertaker: Has Science Buried God?* Lion; p. 182, emphasis and text in square brackets added.

[79] MM/MN = Methodological Materialism / Methodological Naturalism: The assumption that all things can be explained in purely physical or 'natural' terms even if the super natural / the spiritual are real. PM = Philosophical Materialism: The philosophical assumption that only the physical is real.

[80] Dear Reader, please glance at your book's cover. There is no evidence for single cells evolving—randomly or guidedly—into jelly fish or worms, worms into fish, bears into whales, shrews into bats. Just the tired, old assertions that: 'It must be so… because Darwin thought so.'

[81] Lennox, J. 2009. *God's Undertaker.* Lion; p. 143.

[82] Lennox, J. 2009. *God's Undertaker.* Lion; p. 143, emphasis added.

[83] Behe, M. 2008. *The Edge of Evolution.* Free Press

[84] Durston, K. 9.7.15 An Essential Prediction of Darwinian Theory Is Falsified by Information Degradation *Evolution News.* Emphasis added.

[85] Behe, M. 2019. *Darwin Devolves.* Harper One.

[86] Perloff, J. 1999, *Tornado in a Junk Yard* https://jamesperloff.com/why-the-creation-evolution-debate-matters/

[87] Denton, M. 1985. *Evolution: A Theory in Crisis.* Burnett Books. p. 209. Emphasis and text in square brackets added.

[88] Ibid. p. 210–212. Emphasis and text in square brackets added.

[89] Perloff, J. 1999, *Tornado in a Junk Yard* https://jamesperloff.com/why-the-creation-evolution-debate-matters/

[90] Ibid, emphasis, including exclamation marks and text in square brackets added.

[91] Morrison, J. 2019. *Evolution's Final Days: The Mounting Evidence Disproving The Theory of Evolution (problems, myth, hoax, fraud, flaws)* Emphasis added.

[92] Shedinger, R. 2024. *Darwin's Bluff: The Mystery of the Book Darwin Never Finished.* Discovery Institute. Review by Gunter Bechly, on Amazon.

[93] C. Darwin, *On The Origin Of Species*, Emphasis, capitals, bold etc added.

[94] Ibid. p. 261, emphasis added.

95 'In practice, laboratory experiments have shown that right-handed building blocks have a slightly greater affinity, or attraction, for other right-handed building blocks. Therefore, at each step in the addition of another building block, there is a 3/7 chance that the next one added will be the same optical isomer as the one previously added. . . . The smallest known free living life forms, bacteria, have about 12,000,000 nucleotides in their DNA. If we were to calculate the odds of adding twelve million successive right-handed nucleotides to the growing chain, without a single left- handed one being added, it would be 5 raised to the 12 millionth power, (5/12,000,000)!' Eastman & Missler

96 Eastman, M & C. Missler. 1995. *The Creator Beyond Time and Space.* Emphasis, alternative fonts, and text in square brackets added.

97 Steiner, R. 2008 (written 1904–5). *Knowledge of the Higher Worlds and its Attainment.* Tompkins, P. 1997. The Secret Life of Nature, Thorsons

98 Quoted from George Wald, 'The Origin of Life', *Scientific American* 191:48 (May 1954). Emphasis and text in square brackets added.

99 Quoted from George Wald, 'The Origin of Life', *Scientific American 191:48 (May 1954).*

100 uncommondescent.com/intelligent-design/a-world-famous-chemist-tells-the-truth-theres-no-scientist-alive-today-who-understands-macroevolution/#:~:text=Although most scientists leave few,to me, permit unhealthy leeway

101 uncommondescent.com/intelligent-design/a-world-famous-chemist-tells-the-truth-theres-no-scientist-alive-today-who-understands-macroevolution/#:~:text=Although most scientists leave few,to me, permit unhealthy leeway

102 Please see page 243, starting with the last paragraph on 243, for a brief discussion of the problems of 'natural evil,' and suffering on our planet.

103 Michael Flannery. 'Was Darwin a Scholar or a Pitchman?' *Evolution News (October 20, 2015).* Emphasis added.

104 Bethell, T. 2016. *Darwin's House of Cards. Discovery Institute.*

105 Quoted by Lennox, J. 2009. *God's Undertaker.* Lion; p. 143, emphasis added.

106 Latham, A. *The Naked Emperor: Darwinism Exposed.* Janus, p. 88. Text in square brackets added.

107 Wells, J. 2006. *The Politically Incorrect Guide to Darwinism and Intelligent Design.* Regnery.

108 Gee, H. 1999. *In Search of Deep Time: Beyond the Fossil Record to a New History of Life.* The Free Press; pp. 32, 113–17.

109 Davidson, J. 1992. *Natural Creation or Natural Selection – A Complete New Theory of Evolution.* Element; p. 13–14, emphasis and text in square brackets added.

110 Wells, J. 2000. *Icons of Evolution.* Regnery, p. 116.

111 Ibid.

112 Bethell, T. 2016. *Darwin's House of Cards.* Discovery Institute, emphasis and text in square brackets added.

113 Dembski, W. 2007. *No Free Lunch: Why Specified Complexity Cannot Be Purchased without Intelligence.* Rowman Littlefield. 'Very Insightful,' 4. 22.02, amazon review, text in brackets added.

114 Simmons, G. 2007. *Billions of Missing Links.* Harvest House; pps. 207–211. Emphasis added.

[115] Ibid, p. 29. emphasis and text in square brackets added.

[116] Ibid, pps. 207–211,

[117] Behe, M. 1996. *Darwin's Black Box*. The Free Press, p. 185. Emphasis and text in brackets added.

[118] Quoted by Michael E. Tymn, http://whitecrowbooks.com/michaeltymn/biography/ "It wasn't long after the birth of modern *Spiritualism* in 1848 that scientists and scholars began investigating the phenomena. Many of them started out with the intent of showing that all mediums were charlatans, but one by one they came to believe in the reality of mediumship and related psychic phenomena." Tymn,

[119] Flannery, M. A. 2011. *Alfred Russell Wallace: A Rediscovered Life,* Discovery Institute Press

[120] Since the artificial synthesis of urea in 1828 the science view is that there are no Life Forces.

[121] According to science today, all things in Nature are 'coincidences,' and 'purposeless.'

[122] George Wald, 'The Origin of Life', *Scientific American* 191:48 (May 1954). Eastman, M. C. Missler. 1995. *The Creator Beyond Time and Space*

[123] Kelly, E. F. E.W. Kelly, A. Crabtree, A. Gauld 2009 *Irreducible Mind: Toward a Psychology for the 21st Century.*

[124] Behe, M. 1996. *Darwin's Black Box. The Free Press*.

[125] Ibid.

[126] Demolishing Darwin's Tree: Eric Bapteste and the Network of Life. *Evolution News. 9.9.13.*

[127] Gould, S. J. 'Is a new and general theory of evolution emerging?' *Paleobiology, vol. 6(1), January 1980,* p. 127. Emphasis added.

[128] Eldredge, N. & I. Tattersall. 1982. *The Myths of Human Evolution*. Columbia University Press, p. 48.

[129] Simpson, G. G. 1955. *The Major Features of Evolution*. Columbia University Press, p.360.

[130] https://evolutionnews.org/2021/11/the-discontinuous-fossil-record-refutes-darwinian-gradualism/

[131] Battson, A. Conflicts Between Darwin and Paleontology. *http://www.veritas- ucsb.org/library/battson/stasis/2.html*

[132] Gould, S. J. 1980, emphasis added.

[133] C. Darwin, *On The Origin Of Species*, Emphasis, capitals, bold etc added.

[134] Gould, S. J. 1980. *Is a New and General Theory of evolution emerging?* p.120, emphasis added.

[135] Ibid. p. 61 emphasis and words in square brackets added.

[136] Patterson, C. 1978. *Evolution*. The British Museum of Master Books, Natural History.

[137] Sunderland, L. 1988. *Darwin's Enigma*. p. 89. Emphasis and text in square brackets added.

[138] Milton, R. 2000. *Shattering the Myths of Darwinism*. Inner Traditions Bear and Company; back cover, text in brackets added.

[139] Denton, M. 1985. *Evolution: A Theory in Crisis*. Burnett Books. p. 213. Emphasis and text in square brackets added.

¹⁴⁰ Latham, A. 2005. *The Naked Emperor: Darwinism Exposed.* Janus, p. 38; emphasis and words in square brackets added.

¹⁴¹ Simmons, G. 2007. *Billions of Missing Links. Harvest House;* pps. 75–76, emphasis and text in square brackets added.

¹⁴² Darwin, C. 1859 *On the Origin Of Species,* Ch. X, "On the Imperfection of the Geological Record".

¹⁴³ Ibid., Ch. IX, emphasis added.

¹⁴⁴ Darwin, C. 1859. *On the Origin of Species by Means of Natural Selection or the Preservation of Favoured Races in the Struggle for Life,* reprint of 6th edition (John Murray, 1902), p. 341–342. Emphasis added.

¹⁴⁵ Sunderland, L. 1988. *Darwin's Enigma: Ebbing the Tide of Naturalism.* Emphasis added.

¹⁴⁶ *Discover Magazine,* p. 68 (April, 2011).

¹⁴⁷ Margulis, L. & D. Sagan. 2003. *Acquiring Genomes: A Theory of the Origins of the Species.* Basic Books, p. 29. Emphasis and text in square brackets added.

¹⁴⁸ Quoted in "Discover Interview: Lynn Margulis Says She's Not Controversial, She's Right," *Discover Magazine, p. 68 (April, 2011).* Emphasis and text in square brackets added.

¹⁴⁹ Mayr, E. 1942. *Systematics and the Origin of Species.* Dover Publications, p. 296.

¹⁵⁰ Malcolm Muggeridge, *The Advocate,* March 8, 1984, p. 17.

¹⁵¹ Richard Lewontin (1997) Billions and billions of demons (review of The Demon-Haunted World: Science as a Candle in the Dark by Carl Sagan, 1997). *The New York Review, January 9, p. 31.*

¹⁵² Quoted from George Wald, 'The Origin of Life', *Scientific American 191:48 (May 1954).*

¹⁵³ Did a neurosurgeon go to heaven? Why a Near-Death Experience Isn't Proof of Heaven. M. Shermer. *Scientific American. April 13, 2013.*

¹⁵⁴ Carter, C. 2012. *Science and Psychic Phenomena: The Fall of the House of Skeptics.* Inner Traditions.

99 Quoted by Michael E. Tymn, http://whitecrowbooks.com/michaeltymn/biography/ "It wasn't long after the birth of modern Spiritualism in 1848 that scientists and scholars began investigating the phenomena. Many of them started out with the intent of showing that all mediums were charlatans, but one by one they came to believe in the reality of mediumship and related psychic phenomena." Tymn,

¹⁵⁶ Das Wesen der Materie [The Nature of Matter], speech at Florence, Italy (1944) (from Archiv zur Geschichte der Max-Planck-Gesellschaft, Abt. Va, Rep. 11 Planck, Nr. 1797) Emphasis added.

¹⁵⁷ Dawkins, R. 1978. *The Selfish Gene,* Flamingo. Amazon on 40th Anniversary Edition:
"Dawkins articulates a gene's eye view of evolution - a view ... in which organisms can be seen as vehicles for their [the genes] replication."
But no explanation, from Dawkins, as to where the "selfish-genes" came from in the first place? By "pure chance?' Nor does Dawkins tell us, why his 'selfish genes' would ever need to evolve beyond a planet-covering algal or bacterial stage, into bears and whales or caterpillars and butterflies…
Quite apart from the fact that there is NO EVIDENCE that any such process took place.
Those who follow Darwin and Dawkins, it seems to me, see what they want to see, not what is actually there—just like Darwin himself. Darwin's idea of Evolution, Bacterium, to Bear, to Whale, appeals to them, but there is no evidence that such a process ever took place, let alone that it could do so mindlessly and unintelligently, as materialists are obliged to argue—by their philosophy.

[158] Shedinger, R. 2024. *Darwin's Bluff: The Mystery of the Book Darwin Never Finished*. Discovery Institute. The algae idea is mine, but the bacteria reference is Shedinger's.

[159] Denton, M. 1985. *Evolution: A Theory in Crisis*. Burnett Books.

[160] Sunderland, L. 1988. *Darwin's Enigma*. p. 89. Emphasis and text in square brackets added.

[161] Wells, J. 2000. *Icons of Evolution*. Regnery. p. 73.

[162] Ibid. p. 74.

[163] Ibid. p. 71.

[164] Ibid. p. 83.

[165] Ibid. p. 101.

[166] Tim White, quoted in Ann Gibbons, "In Search of the First Hominids," *Science (Feb. 15, 2002), 295:1214–1219.*

[167] For more detailed discussion of the fossil evidence and human origins, see Casey Luskin, "Human Origins and the Fossil Record" in *Science and Human Origins* Discovery Institute Press, 2012, pp. 45–83.

[168] Leslie Aiello, quoted in Richard Leakey and Roger Lewin, Origins Reconsidered: In Search of What Makes Us Human (Anchor Books, 1993), p. 196.

[169] John Hawks et al., "Population Bottlenecks and Pleistocene Human Evolution," *Journal of Molecular Biology and Evolution (2000), 17(1):2–22.*

[170] Ernst Mayr, What Makes Biology Unique? (Cambridge Univ. Press, 2004), p. 198. Emphasis added.

[171] "New study suggests big bang theory of human evolution," (Jan. 10, 2000) at http://www.umich.edu/~newsinfo/Releases/2000/Jan00/r011000b.html. Emphasis added.

[172] Has Science Shown That We Evolved from Ape-like Creatures? Casey Luskin *Salvo magazine September 26, 2013.* Emphasis added.

[173] Language Is a Rock Against Which Evolutionary Theory Wrecks Itself. *Evolution News,* Michael Egnor. 09.19.16. Emphasis and text in brackets added.

[174] Mayr, E. 1942. *Systematics and the Origin of Species. Dover Publications,* p. 296. Text in square brackets added.

[175] Wells, J. 2000. *Icons of Evolution*. Regnery. pp 229 – 230; emphasis and text in brackets added.

[176] Wells, J. 2000. *Icons of Evolution*. Regnery. pp 229 – 230; emphasis and text in brackets added.

[177] uncommondescent.com/intelligent-design/a-world-famous-chemist-tells-the-truth-theres-no-scientist-alive-today-who-understands-macroevolution/#:~:text=Although most scientists leave few,to me, permit unhealthy leeway

[178] Darwin to J.D. Hooker. 2.2.1871. Emphasis and text in square brackets added.

[179] Darwin, C. 1859. *On the Origin of Species.* Emphasis added.

[180] Denton, M. 1985. *Evolution: A Theory in Crisis*. Burnett Books. Text in square brackets added.

[181] Darwin, C. 1872. *On the Origin of Species,* 6th ed, 1962, New York: Collier Books, p. 41.

[182] Darwin, C. 1872 .*On the Origin of Species,* 6th ed, 1962, New York: Collier Books, p. 41.

[183] Denton, M. 1985. *Evolution: A Theory in Crisis*. Burnett Books.

[184] Johnson, P. 2010. *Darwin on Trial.* IVP Books, p. 60, emphasis and text in square brackets added.

[185] Denton, M. 1985. *Evolution: A Theory in Crisis.* Burnett Books.

[186] Eastman, M & C. Missler. 1995. *The Creator Beyond Time and Space.* Emphasis added.

[187] Flew, A. 2009. *There is a God: How the World's Most Notorious Atheist Changed His Mind* Harper One.

[188] The Hottest New Computer Is: DNA *www.evolutionnews.org/2017/10/03*

[189] Richard Dawkins. 2008. *Life: A Gene-Centric View, Craig Venter & Richard Dawkins: A Conversation in Munich. The Edge.* Emphasis added.

[190] Michael Polanyi, "Life's Irreducible Structure," *Science* 160 (1968), 1308–12, quoted by Wells, J. 2006. *Darwinism and Intelligent Design.* Regnery Publishing; pp. 100–101, emphasis added.

See *www.evolutionnews.org*/2017/03/hottest-new-computer-dna

[192] Nagel, T. 2012. *Mind and Cosmos.* Oxford University Press; p.127. Emphasis added.

[193] Ibid. Emphasis added.

[194] Crick F. 2004. *Of Molecules and Men.* Prometheus, p. 10.

[195] The Origin of Life and the Death of Materialism by Stephen C. Meyer, Phd. *The Intercollegiate Review 31, no. 2 (spring 1996).* Text in square brackets added.

[196] Meyer, S. 2009. *Signature in the Cell.* Harper One; p. 405, emphasis added.

[197] Ibid. p. 405, emphasis and text in square brackets added.

[198] Crick F. 1995. *The Astonishing Hypothesis.* Scribner, p. 3, text in square brackets added.

[199] Denton, M. *1985. Evolution a Theory in Crisis.* Burnet Books, p.250.

[200] Schwartz, G.E. Phd. W.L. Simon. 2002. *The Afterlife Experiments: Breakthrough Scientific Evidence of Life After Death.* Atria Books. Fontana, D. 2005. *Is there an Afterlife: A Comprehensive Overview of the Evidence.* O Books.

[201] MM/MN Methodological Materialism and Methodological Naturalism are methods in science. PM/PN Philosophical Materialism and Philosophical Naturalism are ideologies/belief systems.

[202] Eastman, M & C. Missler. 1995. *The Creator Beyond Time and Space.* Emphasis added.

[203] Nagel, T. 2012. *Mind and Cosmos: Why the Materialist Neo-Darwinian Conception of Nature is Almost Certainly False.* Oxford University Press.

[204] Kelly, E. F. E.W. Kelly, A. Crabtree, A. Gauld 2009 *Irreducible Mind: Toward a Psychology for the 21st Century.* Rowman and Littlefield, product description, emphasis added.

[205] Tallis, R. 2011. *Aping Mankind: Neuromania, Darwinitis and the Misrepresentation of Humanity.* Acumen Publishing; product description.

[206] Carter, C. 2010. *Science and the Near Death Experience.* Inner Traditions; p. 74. Emphasis and text in square brackets added.

[207] Searle, *The Rediscovery of the Mind,* 3–4. Emphasis added.

[208] Radin, D. 1997. *The Conscious Universe: The Scientific Truth of Psychic Phenomena.* Harper Collins

[209] Carter, C. 2010. *Science and the Near Death Experience.* Inner Traditions; p.235.

[210] Gustus, S. 2010. *Less Incomplete: A Guide to Experiencing the Human Condition Beyond the Physical Body.* O Books. Monroe, R.A. 1972. *Journeys Out of the Body.* Souvenir Press. Buhlman, W. 2001. *The Secret of the Soul: Using Out-of-Body Experiences to Understand Our True Nature.* Harper One. Peake, A. 2011. *The Out of Body Experience: The History and Science of Astral Travel.* Watkins Publishing.

[211] Schwartz, G.E. Phd. W.L. Simon. 2002. *The Afterlife Experiments: Breakthrough Scientific Evidence of Life After Death.* Atria Books.

[212] *www.subversivethinking.blogspot.co.uk/2011/05/interview-with-philosopher-* neal.html Emphasis and text in square brackets added.

[213] Carter, C. 2010. *Science and the Near Death Experience.* Inner Traditions – Quoted in Koestler, The Roots of Coincidence, 77.

[214] Crick F. 2004. *Of Molecules and Men.* Prometheus, p. 10. Text in square brackets added.

[215] Morowitz, *"Rediscovering the Mind,"* 12. Emphasis and text in square brackets added.

[216] Carter, C. 2010. *Science and the Near Death Experience.* Inner Traditions; p. 74. emphasis added.

[217] Rao, *"Consciousness."* Emphasis added.

[218] Carter, C. 2010. *Science and the Near Death Experience.* Inner Traditions; p. 75. emphasis added.

[219] Rosenblum and Kuttner, *"Consciousness and Quantum Mechanics: The Connection and Analogies,"* 248, emphasis and text in square brackets added.

[220] Dennett D. 1992. *Consciousness Explained.* Back Bay Books, p. 406.

[221] Richard Lewontin (1997) Billions and billions of demons (review of The Demon-Haunted World: Science as a Candle in the Dark by Carl Sagan, 1997). *The New York Review, January 9, p. 31.*

[222] Paley, W. 1802. *Natural Theology*

[223] Gates, B. 1996. *The Road Ahead.* Penguin. p228.

[224] Eastman, M C. Missler. 1995. *The Creator Beyond Time and Space.*

[225] Moore, E. A. 11.17.2010. cnet.

[226] Quoted from George Wald, 'The Origin of Life', *Scientific American* 191:48 (May 1954).

[227] Greyson, B.: "Commentary on 'Psychophysiological and Cultural correlates Undermining a Survivalist Interpretation of Near-Death Experiences.'" *Journal of Near Death Studies* 26, no. 2 (Winter 2007); P.142.

[228] Crick F. 2004. *Of Molecules and Men.* Prometheus, p. 10. Text in square brackets added.

[229] Wells, J. 2006. *The Politically Incorrect Guide to Darwinism and Intelligent Design.* Regnery Publishing.

[230] *Can You Tell Me Anything About Evolution?* Presentation by Colin Patterson at the American Museum of Natural History. Nov. 1981. Emphasis added. A CD and annotated transcription of the talk, including Patterson's interactions with luminaries like Niles Eldridge in the Q&A and lively discussion that followed the talk was, at one time, available at www.arn.org.

[231] Ibid. Emphasis added. First saw Colin Patterson's famous comments on 'evolution as faith' rather than genuine scientific knowledge, in Phillip Johnson's *Darwin on Trial.*

[232] Monod, J. 1972. *Chance and Necessity.* Collins. pp 134–135.

[233] The 'Wow! signal' of the terrestrial genetic code. *j.icarus.* 2013.02.17

[234] Ibid.

[235] *Evolution News* review of the "Wow!" Signal of Intelligent Design published in the Planetary Science Journal Icarus. *www.evolutionnews.org.* Text in square brackets and emphasis added.

[236] Bruce Alberts. 'The cell as a collection of protein machines: Preparing the next generation of molecular biologists'. *Cell,* 92 (February 8, 1998): 291.

[237] Goethe's Mephistopheles tricked Faust into selling his own Soul. The word "could derive from the Hebrew mephitz, meaning "destroyer", and tophel, meaning "liar"... The name can also be a combination of three Greek words: "me" as a negation, "phos" meaning light, and "philis" meaning loving, making it mean "not-light-loving..."' *wikipedia.*

[238] Carter: "For the materialist, the term *unscientific* seems to be the modern equivalent of the term heretical, and it is [used] to exclude from consideration ideas that challenge the believer's faith."

[239] Carter, C. 2010. *Science and the Near Death Experience.* Inner Traditions; p.238. Text in square brackets added.

[240] *Ibid, p.237.*

[241] Crick F. 1995. *The Astonishing Hypothesis.* Scribner, p. 3, text in square brackets added.

[242] Lesiola, M. 2018. *Heretic: One Scientist's Journey from Darwin to Design*, Discovery Institute Press. Valtaoja said to Leisola: "Life is nothing else than physics and chemistry—mere electricity. There is no reason to assume anything supernatural."

[243] *Ibid. p 328.*

[244] Das Wesen der Materie [The Nature of Matter], speech at Florence, Italy (1944) (from Archiv zur Geschichte der Max-Planck-Gesellschaft, Abt. Va, Rep. 11 Planck, Nr. 1797) Emphasis added.

[245] J. Jeans, 1931. *The Mysterious Universe.* Cambridge University press, p. 137.

[246] Ibid

[247] Wilberg, P. 2008. *The Science Delusion: Why God is Real and 'Science' is Religious Myth.* New Gnosis Publications. P. 141; emphasis and text in square brackets added.

[248] Nagel, T. 2012. *Mind and Cosmos: Why the Materialist Neo-Darwinian Conception of Nature is Almost Certainly False.* Oxford University Press.

[249] Lovelock, J. 1979. *Gaia: A New Look at Life on Earth.* Oxford University Press. Lovelock, J. 2010. *The Vanishing Face of Gaia: A Final Warning.* Penguin

[250] Brinkley, D. & K. Brinkley. 2012. *Secrets of the Light: The incredible true story of one man's near-death experiences and the lessons he received from the other side.* Piatkus.

[251] Steiner, R. 1918. *Knowledge of Higher Worlds and its Attainment.* Rudolf Steiner Press. 6th Ed. 2004.

[252] Besant, A. Leadbeater, C.W. C. Jinarajadasa. 1908. *Occult Chemistry.* Theosophical Publishing House

[253] Tompkins, P. 1997. *The Secret Life of Nature*, Thorsons, pp. 68–69. Emphasis and square brackets added.

254 Philips, S.M. 1980. *Extra–Sensory Perception of Quarks*. Theosophical Publishing.

255 *Kindred Spirit Magazine*. 1998, June, text in square brackets added.

256 Tompkins, P and C. Bird. 1973. *The Secret Life of Plants*. Harper Perennial

257 Rudolf Steiner: "Steiner's gift to the world was a moral and meditative way to objective vision... If accepted... it could bridge the existing cleft between a man's religious conviction and his intellect and will." Franz Winkler, M.D., *Man the Bridge between Two Worlds*. "That the academic world has managed to dismiss Steiner's works as inconsequential and irrelevant, is one of the intellectual wonders of the twentieth century. Anyone who is willing to study those vast works with an open mind (let us say, a hundred of his titles) **will find himself faced with one of the greatest thinkers of all time**, whose grasp of the modern sciences is equaled only by his profound learning in the ancient ones." Russell W. Davenport, *The Dignity of Man*.' Amazon.

258 Taylor, S. 2010. *Waking From Sleep*. Hay House, p. 44.

259 'Objective Matter' which, essentially, is not '**solid**' at all, but is made of S....P....A....C....E....

260 Capra, F. 1992. *The Tao of Physics*. Flamingo. 'The... classic exploration of the connections between Eastern mysticism and modern physics. An international bestseller, the book's central thesis, that the **mystical traditions of the East constitute a coherent philosophical framework within which the most advanced Western theories of the physical world can be accommodated**, has not only withstood the test of time but is ever more emphatically endorsed by ongoing experimentation and research.'

261 For example Christian mysticism, Gnosticism, Shamanism, Theosophy, Anthroposophy, Kabala, Sufism, Yogism and Taoism

262 Rupert Sheldrake's Website: Dialogues and Controversies; emphasis and text in brackets added.

263 **MM/MN** – methodological materialism /methodological naturalism, **PM/PN** – philosophical materialism / philosophical naturalism.

264 Schwartz, G.E. Phd. W.L. Simon. 2002. The Afterlife Experiments: Breakthrough Scientific Evidence of Life After Death. Atria Books.

265 Zammit, V. W. 2013. *A Lawyer Presents Evidence for the Afterlife*. White Crow Books.

266 Carter, C. 2010. *Science and the Near Death Experience*. Inner Traditions; p.237.

267 Carter, C. 2010. *Science and the Near Death Experience*. Inner Traditions, amazon, T. Skaftnesmo, text in square brackets added.

268 Neal, M. 2012. *To Heaven and Back: A Doctor's Extraordinary Account of Her Death, Heaven, Angels, and Life Again: A True Story*. Waterbrook Press. amazon interview.

269 Ibid.

270 Nelson, B. 2007. *The Emotion Code: How to Release Your Trapped Emotions for Abundant Health, Love and Happiness*. Wellness Unmasked Publishing; pp. 268–274.

271 Described, by her, also on youtube, https://www.youtube.com/watch?v=LOhCdfIip10

272 Ibid, p. 69.

273 Ibid, pp. 69–70, emphasis added.

274 Atwater, P.M.H. 1994. *Beyond the Light: Near Death Experiences*. Thorsons; p. 28, emphasis added.

275 Lommel, P. 2010. *Consciousness Beyond Life, The Science of the Near Death Experience*. Harper One

276 Carter, C. 2010. *Science and the Near Death Experience*. Inner Traditions.

277 Ibid, p.235.

278 Schwartz, G.E. Phd. W.L. Simon. 2002. *The Afterlife Experiments: Breakthrough Scientific Evidence of Life After Death*. Atria Books.

279 Grossman, N. 2002. *Who's Afraid of Life After Death?* p.8

280 Carter, C. 2010. *Science and the Near Death Experience*. Inner Traditions; p.237, emphasis added.

281 Dossey, L. 1997. *Recovering the Soul: A Scientific and Spiritual Search*. Bantam.

282 Stevenson, I. 1980. *Twenty Cases Suggestive of Reincarnation: Second Edition, Revised and Enlarged*. University of Virginia Press. Stemman, R. 2012. *The Big Book of Reincarnation: Examining the Evidence that We Have All Lived Before*. Hierophant Publishing.

283 Newton, M. 1994. *Journey of Souls: Case Studies of Life Between Lives*. Lewellyn Publications. Over a thirty year period Newton regressed 7,000 clients, building up a picture of the inter-life as his clients, in hypnotic regression, reported on the same themes again and again: including assessment of the completed lives, further experiences in the spiritual dimensions and eventual return to physical life.

284 Gustus, S. 2011. *Less Incomplete: A Guide to Experiencing the Human Condition Beyond the Physical Body*. O Books. Emphasis added.

285 Tompkins, P. 1997. *The Secret Life of Nature,* Thorsons, pp. 68–69.

286 Kelly, Kelly, Crabtree & Gauld 2009 *Irreducible Mind: Toward a Psychology for the 21st Century*. Rowman and Littlefield.

287 Zammit, victorzammit.com

288 Carter, C. 2012. *Science and the Afterlife Experience*. Inner Traditions, review, amazon.

289 Ibid.

290 Radin, D. 2009. *The Conscious Universe: The Scientific Truth of Psychic Phenomena.* Harper One, Radin's book contains much documentation of this kind of evidence.

291 McLuhan, R. 2010. *Randi's Prize: What Sceptics Say About the Paranormal, Why They Are Wrong, and Why It Matters*. Matador. Amazon review: Geophysics Ph.D, Sun D, text in square brackets added.

292 In Hinduism, the sacred syllables A-U-M represent the primal sound and vibration of God or It,Is,I,S/He,You,All, ISHYA, giving creative expression to the universe—from no-thing to every-thing.

293 There is a new science called biomimetics, devoted to learning from and to imitating systems in nature for the purpose of solving complex engineering and other design problems.

294 Brinkley, D. 1994. *Saved By The Light: The incredible Story of a Man Who Died Twice and The Prophetic Revelations He Received*. Piatkus. This was one of the potential futures he was shown.

295 James, W. 1902. *The Varieties of Religious Experience*. Penguin Classics.

296 Sheldrake, R. 2012. *The Science Delusion*. Coronet; product description, amazon.

297 Mary Midgley*, The Guardian;* emphasis and text in square brackets added.

298 Tompkins, P C. Bird. 1989. *The Secret Life of Plants*. Harper Perennial.

299 Atwater, P.M.H. 1994. *Beyond the Light: Near Death Experiences*. Thorsons; p. 28.

300 Nelson, B. 2007. *The Emotion Code: How to Release Your Trapped Emotions for Abundant Health, Love and Happiness*. Vermillion 2019

301 Zeitgeist Conference, by Google, 2011. *Hawking said that "philosophy is dead."* Wikipedia.

www.ingramcontent.com/pod-product-compliance
Ingram Content Group UK Ltd.
Pitfield, Milton Keynes, MK11 3LW, UK
UKHW022350210325
5111UKWH00008B/220